Transmission Principles for Technicians

Transmission Principles for Technicians

Second Edition

D C Green

M Tech, CEng, MIEE

Longman
Scientific &
Technical

Longman Scientific & Technical
Longman Group UK Limited
Longman House, Burnt Mill, Harlow
Essex, CM20 2JE, England
and Associated Companies throughout the world

Copublished in the United States with
John Wiley & Sons, Inc., 605 Third Avenue, New York
NY 10158

First published 1988
Second impression 1991
Third impression 1992
Second edition 1995

British Library Cataloguing in Publication Data
A catalogue entry for this title is available from the British Library

ISBN 0-582-24515-X

Transferred to digital print on demand 2002

Printed and bound in Great Britain by Antony Rowe Ltd, Eastbourne

Contents

Preface

This book provides an introduction to the basic principles and practice of telecommunication line systems. The intended readership in the United Kingdom is the student of either the Business and Technician Education Council (BTEC) or the City and Guilds of London Institute units in Transmission Principles at both levels II and III.

A study of the book should give the reader the understanding of the operation of both analogue and digital transmission systems that is necessary for all telecommunication technicians, whether their particular interests lie in the field of data, line or radio communications, or in that of switching systems.

The earlier chapters of the book present, using a minimum of mathematics, the basic concepts of signals, modulation, noise and transmission lines. These concepts are then employed, in the later chapters, to introduce the reader to practical transmission systems.

Many worked examples are provided in the text to illustrate the principles that have been discussed. At the end of each chapter will be found a number of exercises for the reader to check understanding of each topic. Answers to numerical exercises are given at the end of the book.

D.C.G.

1 Signals

In a telecommunication system, media are provided for the transmission of intelligence from one point to another. For example, telephone conversations and data signals are transmitted over both copper and optical fibre cables and radio signals are transmitted through the atmosphere. It is necessary to ensure that sufficient information is available at the receiving end of a system to allow the received intelligence to be understood and appreciated by the person, or equipment, receiving it. The requirements demanded of the transmission media depend upon the type of intelligence to be transmitted, but for the transmission of speech, music and television signals the main requirement is that sufficient frequencies are retained in the transmitted waveform to permit the received sound and picture to be intelligible and, in the case of music and television, to be enjoyable also. It is therefore necessary to have an appreciation of the range of frequencies produced by the human voice, and by musical instruments, and the frequency range over which the human ear is capable of responding. For the picture signal in a television system and for data signals it is also necessary that the waveform of the received signal bears sufficient resemblance to the original waveform to allow the original information to be re-created at the receiver.

The Voice and Speech

A current of air expelled by the lungs passes through a narrow slit between the vocal cords in the larynx and causes them to vibrate. This vibration is then communicated to the air via various cavities in the mouth, throat and nose. The shape and size of the nose cavities of each individual are more or less fixed, but the mouth and throat cavities can have their shapes and dimensions considerably changed by the action of the tongue, lips, teeth and the throat muscles. The frequency at which a particular cavity allows the air inside the mouth to vibrate most freely depends upon the shape and dimensions of the cavity; these can readily be adjusted by the movement of the lips, tongue and teeth. The pitch of the spoken sound depends upon both the length and the tension of the vocal cords and the width of the slit between them. The length of the vocal cords varies from person to person; for example, a woman will have shorter vocal cords than a man, while the tension and distance apart of the vocal cords are under the control of muscles.

When a person speaks, his vocal cords vibrate and the resulting sounds, which are rich in harmonics but of almost constant pitch, are carried to the cavities in the mouth, throat and nose. Here the sounds are given some of the characteristics of the desired speech by the emphasizing of some of the harmonics contained in the sound waveforms, and by the suppression of others. Sounds produced in this way are the vowels, a, e, i, o and u, and these contain a relatively large amount of sound energy. Consonants are spoken with the lips, tongue and teeth and they contain much smaller amounts of energy and often include some relatively high frequencies.

The sounds produced in speech contain frequencies which lie within the frequency band 100–10 000 Hz. The pitch of the voice is determined by the fundamental frequency of the vocal cords and this is about 200–1000 Hz for women and about 100–500 Hz for men. The tonal quality and the individuality of each person's voice are determined by the higher frequencies that are produced.

The power content of speech is small, a good average being of the order of 10–20 microwatt. However, this power is not evenly distributed over the speech frequency range, most of the power being contained at frequencies in the region of 500 Hz for men and 800 Hz for women.

Music

The notes produced by musical instruments occupy a much larger frequency band than that occupied by speech. Some instruments, such as the organ and the drum, have a fundamental frequency of 50 Hz or less while many other instruments, for example the violin and the clarinet, can produce notes having a harmonic content in excess of 15 000 Hz. The power content of music can be quite large. A sizeable orchestra may generate a peak power somewhere in the region of 90–100 watts while a bass drum well thumped may produce a peak power of about 24 watts.

Hearing

When sound waves are incident upon the ear they cause the ear drum to vibrate. Coupled to the ear drum are three small bones which transfer the vibration to a fluid contained within a part of the inner ear known as the cochlea. Inside the cochlea are a number of hair cells and the nerve fibres of these are activated by vibration of the fluid. Activation of these nerve fibres causes them to send signals, in the form of minute electric currents, to the brain where they are interpreted as sound.

The ear can only hear sounds whose intensity lies within certain limits; if a sound is too quiet it is not heard and, conversely, if a sound is too loud it is felt rather than heard and may cause discomfort or even pain. The minimum sound intensity, measured in pascals (1 Pa = 1 Nm^2), that can be detected by the ear is known as the *threshold of hearing or audibility* and the sound intensity that just produces a feeling of discomfort is known as the *threshold of feeling*. The ear is not, however, equally sensitive at all frequencies, as shown by Fig. 1.1. In this figure curves have been plotted showing how the thresholds of audibility and feeling vary with frequency for the average person.

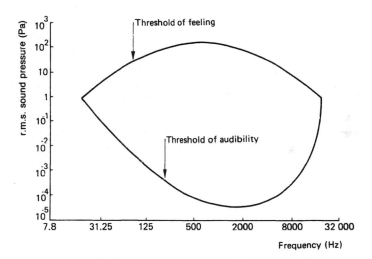

Fig. 1.1 The thresholds of audibility and feeling

It can be seen that the frequency range over which the average human ear is capable of responding is approximately 30–16 500 Hz, but this range varies considerably with the individual. The ear is most sensitive in the region of 1000 to 2000 Hz and it becomes rapidly less sensitive as the upper and lower limits of audibility are approached. The limits of audibility are clearly determined not only by the frequency of the sound but also by its intensity. See, for example, the increase in the audible frequency range when the sound intensity is increased from, say, 1×10^{-3} Pa to 1×10^{-2} Pa. At the upper and lower limits of audibility the thresholds of audibility and feeling coincide and it becomes difficult for a listener to distinguish between hearing and feeling a sound.

The Transmission of Sound over a Simple Telephone Circuit

The arrangement of a simple, unidirectional telephone circuit is shown in Fig. 1.2.

The carbon granule telephone transmitter is a transducer whose electrical resistance varies in accordance with the waveform of the sound incident upon it. The telephone receiver is a transducer whose

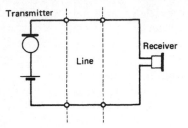

Fig. 1.2 A simple unidirectional speech circuit

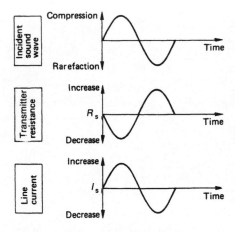

Fig. 1.3 The relationship between the sound incident on a transmitter, the microphone resistance, and the current flowing to line. R_s = microphone resistance when no sound is incident upon it. I_s = d.c. current to line when no sound is incident on the microphone

diaphragm vibrates when a varying current is passed through it; these vibrations produce sound waves having the same waveform as the current.

When no sound is incident upon the transmitter, the resistance of the transmitter is constant and a d.c. current flows into the line from the battery. When a person speaks into the transmitter, its resistance is caused to vary in the same way as the speech waveform. This, in turn, produces variations in the current flowing to line. For example, during one half-cycle of the speech waveform the resistance of the transmitter is decreased and so the line current is increased, while during the next half-cycle the transmitter resistance is increased and the line current is decreased. Thus the line current is continuously varying about its d.c. value as shown by Fig. 1.3. The varying line current passes through the receiver at the distant end of the line and causes the receiver's diaphragm to vibrate and reproduce the original speech.

For a two-way conversation this simple arrangement would have to be duplicated, the second circuit having the positions of the transmitter and battery and of the receiver reversed. Such an arrangement would be uneconomic since it would require two pairs in a telephone cable, and telephone cables are very expensive. A further disadvantage of the simple circuit is that as the length of line is increased, the variations in the resistance of the transmitter become an increasingly small fraction of the line resistance. This means that the magnitude of the a.c. component of the line current decreases with increase in line length until the a.c. current is not large enough to operate the receiver.

A simple telephone circuit which overcomes these disadvantages is shown in Fig. 1.4. In this circuit, each transmitter is connected to the line via a transformer but the two receivers are directly connected to line. When a person speaks into either of the two transmitters, the resulting changes in transmitter resistance cause a relatively large a.c. current to flow in the local transmitter circuit. In passing through the transformer primary winding, this a.c. current induces an e.m.f. into the secondary winding. The induced e.m.f. drives an a.c. current, having the same waveform as the speech, to line. At the distant end of the line the a.c. current flows in the receiver and causes its diaphragm to vibrate and so produce sound waves which are similar to the original speech.

The circuit given in Fig. 1.4 has the disadvantage of requiring a separate battery at each telephone. The vast majority of telephone circuits are connected to a local telephone exchange and these circuits

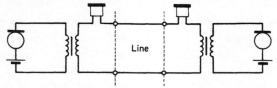

Fig. 1.4 A simple two-way speech circuit

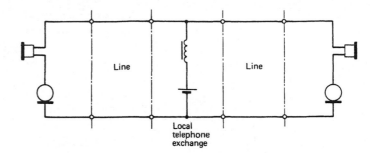

Fig. 1.5 The central battery system

are all operated from a *central battery*. The basic principle of the central battery system is illustrated by Fig. 1.5. A large capacity secondary cell battery is installed at the local telephone exchange and supplies current to the lines when the telephone handsets are lifted from their rests. When either of the transmitters is spoken into, a speech-frequency current is superimposed upon the battery current and is transmitted, via the telephone exchange, to the other telephone. The inductor is connected in series with the exchange battery to prevent the speech-frequency currents entering the battery. In practice, a more complex telephone circuit is needed because the circuit as shown would suffer from the speech-frequency currents generated by the transmitter flowing in the associated receiver. The sound picked up by the transmitter would be clearly audible in the receiver to give excessive sidetone.

Sidetone makes it difficult, if not impossible, to carry on a conversation and so it must be reduced. Sidetone is never completely eliminated because if it were the telephone would appear to be 'dead'. The block diagram of a modern electronic telephone is shown by Fig. 1.6. The telephone has three main functions: (*a*) the transmission and

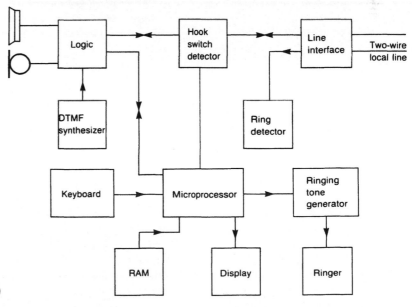

Fig. 1.6 Electronic telephone
(DTMF: dual-tone multi-frequency)

reception of speech signals, (b) the transmission of routing digits and (c) ringing to indicate that an incoming call has arrived. These functions can now be performed by a single integrated circuit (IC) but often separate ICs are used.

Telephone Networks

The telephone network of the UK is divided into the access and core networks. In the access network local lines connect the individual telephones, FAX machines or data terminals to their local telephone exchange. In the core network trunk circuits interconnect telephone exchanges, both local and trunk switching, to one another.

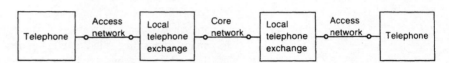

Fig. 1.7 Telephone circuit

The basic arrangement employed is shown by Fig. 1.7. The telephone is connected by copper conductors in the access network to the local telephone exchange. Here the analogue signal is converted into digital form by an *analogue-to-digital converter* (ADC) before being applied to the switching and routing equipment in the core network. The signal is then transmitted over the digital trunk network, passing over any combination of coaxial cable, optical fibre cable and microwave radio system, to reach the destination local telephone exchange. At this point the signal is routed to the wanted local telephone line and converted back into its original analogue form by a *digital-to-analogue converter* (DAC). The eventual aim is to bring the digital network directly into the customer's premises as shown by Fig. 15.1 (see page 216).

Long-distance telephone lines are extremely expensive and it is not economically possible to connect every exchange in the network to every other exchange; direct trunk circuits are only provided between two exchanges when justified by the traffic carried. The vast majority of trunk circuits are routed over one or more multi-channel telephony systems. International circuits may be routed over submarine cable, microwave radio or satellite multi-channel systems, or sometimes over high-frequency (3−30 MHz) radio links.

Television Signals

Television is the transmission of moving pictures. To give the impression of movement a series of still pictures is shown one after the other, each picture being slightly different from the preceding one. As long as the pictures are shown quickly enough the 'persistence of vision' of the human eye makes the viewers believe that they are seeing a moving picture.

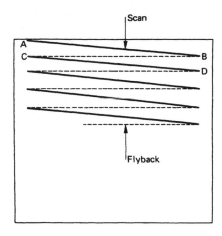

Fig. 1.8 Sequential scanning of a television picture

A television picture is divided vertically into 625 lines, each of which is effectively subdivided into a number of elemental areas known as *pixels*. If the pixels are small enough each will have a constant brightness; for monochrome television, the brightest areas are white, the darkest black, and all other areas are various shades of grey. At the television studio a television camera repeatedly scans the scene to be televised. As each pixel is scanned a voltage that is directly proportional to the brightness of the scene is generated. The action of the television camera is to convert the brightness of each area into a voltage whose amplitude is proportional to that brightness.

The camera is unable to transmit all the voltages simultaneously, and so the system is arranged to transmit them sequentially. The picture is scanned by the camera in a series of lines as in Fig. 1.8. The scan starts at the top left-hand corner of the picture, point A, and travels along the first line to point B. As each elemental area is scanned a proportional instantaneous voltage is generated and a voltage waveform representing the variation of brightness along the first line is produced. At the end of this line the scan flies rapidly back to point C before travelling along the second line to point D, and so on until the entire picture has been scanned. When the end of the last line has been reached the scan is moved back to point A ready to commence scanning the next field.

The detail of the light image to be transmitted is represented by a voltage having a maximum amplitude of $+0.7$ V, indicating peak white, and a minimum amplitude of 0 V, indicating black. The analogue waveform is known as the *picture signal*. At the television receiver the picture is built up from a number of lines that are traced out in sequence by a spot of light travelling on the screen of the cathode-ray tube. The spot is produced by an electron beam incident upon the inner face of the screen, and to obtain the various shades of grey, and black and white needed to create a particular picture the brightness of the spot is modulated by the picture signal. For the picture to be reproduced correctly it is essential for the two scans (camera and receiver) to be in synchronism. The picture signal produced by the television camera is therefore accompanied by *synchronizing pulses*. The combination of the picture signal and the synchronizing pulses is known as the *video signal*. A typical video waveform is shown in Fig. 1.9.

The persistence of vision of the human eye enables the viewer to see a complete picture and not the moving spot of light. The fields must be repeated at a frequency high enough, usually 50 Hz, for the slight changes in successive fields to give the impression of movement without apparent flicker. The necessary field frequency is reduced by 50 per cent since *interlaced scanning* is used; here the odd lines, 1, 3, 5, 7, etc., are scanned first and the even lines afterwards. Each complete picture consists of two fields, one field made up using the odd-numbered lines, 1, 3, 5, etc., and the other field using the even-numbered lines, 2, 4, 6, etc. In the UK system 25 pictures are scanned

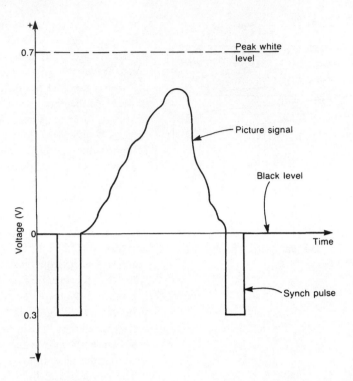

Fig. 1.9 Video signal

every second and this means that the line frequency is 25 $\times$ 625 = 15 625 Hz.

The bandwidth required for a television picture signal depends upon a number of factors, such as the number of lines that make up the picture, the number of fields transmitted per second and the durations of the synchronizing pulses. In practice, the normal bandwidth provided is 5.5 MHz.

A colour television picture signal is transmitted in two parts: (*a*) a brightness (*luminance*) signal, which corresponds to the monochrome signal previously described, and (*b*) colour (*chrominance*) information which is transmitted as the amplitude-modulation sidebands of two colour *subcarriers* which are of the same frequency (approximately 4.434 MHz) but are 90° apart. No extra bandwidth is needed to accommodate the colour information. The luminance signal is used to control the brightness of the colour picture and it also produces the picture displayed by a monochrome receiver. The chrominance signal controls both the hue and the saturation of the colours in the colour picture. A monochrome receiver does not respond to the chrominance signal.

Telegraph Signals

Telegraphy is the passing of messages by means of a signalling code such as the Morse code and the Murray code.

In the *Morse code* characters are represented by a combination of *dot* signals and *dash* signals; the difference between a dash and a dot

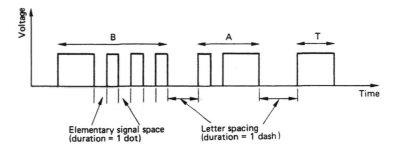

Fig. 1.10 BAT in Morse code

is one of time duration only, a dash having a period three times that of a dot. Spacings between elementary signals, between letters and between words are also distinguished from one another by different time durations. As an example, Fig. 1.10 shows the word BAT in Morse code. The Morse code is not convenient for use with automatic-printing receiving equipment because the number of signal elements needed to indicate a character is not the same for all elements, and the signal elements themselves are of different lengths. Morse code is rarely used in modern telegraphy systems.

In the *Murray code*, or *International Alphabet No. 2* (IA2), all characters have exactly the same number of signal elements and all the signal elements are of the same length. Each character is represented by a combination of five signal elements that may be either a *mark* or a *space*. A mark is represented by either a negative potential or the presence of a tone, and a space is represented by either a positive potential or the absence of a tone. Figure 1.11 shows the letters B, T, R and Y in IA2 code. The IA2 code is used for teleprinter systems.

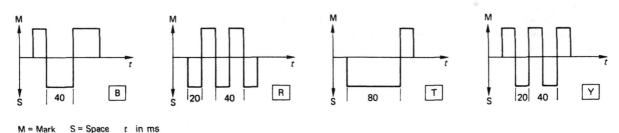

M = Mark   S = Space t in ms

Fig. 1.11 B, T, R and Y in Murray code

Telegraph speed is measured in terms of a unit known as the *baud*. The baud speed of a telegraph signal is the reciprocal of the time duration of the shortest signal element employed. In the Morse code the dot is the shortest signal element; in the IA2 code the elements are all of the same length. The bandwidth required for the transmission of a signal in Morse code depends upon the number of words transmitted per minute and has not been standardized, but it generally lies in the range 100–1000 Hz. Teleprinters are normally operated at a speed of 50 bauds, and this means that the time duration of a mark, or of a space, is 1/50 second, or 20 ms. The bandwidth required to transmit a teleprinter signal depends upon the characters sent, but the maximum bandwidth is demanded when alternate marks and spaces are transmitted, i.e. the letters R and Y. The periodic time of one cycle of the waveforms for R and Y is 40 ms, and so the fundamental frequency of the waveform is 1000/40, or 25 Hz.

Telex

The Telex system has been in existence for many years for the transmission of printed information from one point to another. The system used to employ teleprinters in the customer's premises but today modems are employed to link the customer's computer equipment to the nearest Telex exchange. The customer's equipment is a special-purpose microcomputer that is provided with a visual display, with text-editing software, and with floppy disk storage. Often a PC is provided with a Telex card and it is then able to operate as a Telex terminal.

Both the Telex terminal and the Telex exchange are microprocessor controlled and hence they use the ASCII code internally. The terminals usually are configured to transmit text to line using the IA2 (Murray) code, although they can often also transmit and receive ASCII. The usual transmission speed is 50 bits/s but up to 300 bits/s is common.

Telex is the only established world-wide system available for the transmission of printed messages. Its usefulness is limited by its low transmission speed and its limited character set and it is increasingly being replaced by *facsimile telegraphy* (FAX).

Facsimile Telegraphy

Facsimile telegraphy or FAX is the transmission of photographs, diagrams, documents, etc. over the PSTN. FAX terminals are classified by the ITU−T (International Telecommunication Union−Telecommunication Sector) into four groups.

Group 1: transmission of an A4 document in 6 minutes.
Group 2: transmission of an A4 document in 3 minutes.
Group 3: transmission of an A4 document in under 1 minute.
Group 4: transmission of a document over a digital network.

Currently, the majority of terminals in use are group 3 types and groups 1 and 2 are obsolete. All FAX terminals, other than those in group 4, are designed to work over the analogue telephone network.

The basic block diagram of a group 3 FAX machine is shown in Fig. 1 12. The document to be transmitted is passed in front of the scanner, which scans it line-by-line. Each line consists of 1728 *picture elements*, or PELs, and each PEL is quantized into either black or white. The quantized signal is first coded into digital form and then compressed to remove all redundant information. The coded and

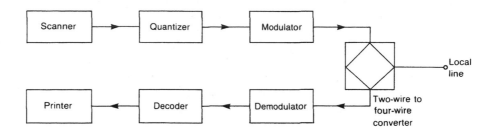

Fig. 1.12 FAX machine

compressed signal is applied to a digital modulator and sent to line. Call control messages are sent at 300 bits/s using ITU−T V21* frequency shift modulation; V29 modulation is used for the transmission of the FAX information at either 9600 bits/s or 7200 bits/s. If the connection via the public switched telephone network (PSTN) is poor then the signalling method will change to V27 ter at either 4800 bits/s and 2400 bits/s.

There are 1145 lines in each page of information and hence $1145 \times 1728 = 1.9786 \times 10^6$ bits make up each page. At one bit per PEL this would take $(1.9786 \times 10^6)/9600 = 206$ seconds to transmit. Fortunately, the scanned data contains redundant information and so it can be compressed by a factor of about ten; this results in each page being transmitted in under one minute.

An incoming message is demodulated, expanded and decoded before it is printed out by the printer. The printer is a thermal, or an ink jet, or a laser type, similar to those employed in conjunction with PCs.

Group 4 FAX is intended for use over 64 kbits/s private digital circuits, or an integrated service digital network (ISDN) local link and the digital trunk network. There are very few such systems in operation at present.

Data Signals

Both mainframe and mini digital computers are widely used by many organizations for commercial data processing applications, such as the calculation and addressing of bills and the compilation of company statistics and records. Much of the data to be processed and perhaps stored by a computer is originated at, and the results are required by, offices that are not located at the same geographical point as the host computer. This means that there is a need for data links that will enable branch offices to communicate with a central host computer. Other examples of the use of data links are electronic mail, the checking of credit cards, and bank and building society cash dispenser points.

A data terminal may consist of a monitor and a keyboard or it may be a micro, or personal (PC), computer. An example of a typical data system is shown by Fig. 1.13. The main office of a mail-order firm has a large number of terminals each of which consists of a monitor and a keyboard. The terminals are all connected to a host computer (mainframe or mini). Several printers, hard disk stores and backing

* See Table 4.1 for ITU−T V series recommendations.

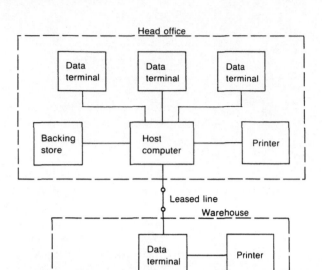

Fig. 1.13 Data system

storage are also connected to the computer. The computer is linked by a leased, permanently set up, connection via the PSTN to a data terminal in the distant warehouse at which the goods are kept. The computer's backing storage contains details of the products in the catalogue, their current prices, the stocks held, and customer's names and addresses. The staff receiving orders over the telephone have access to all this data. When an order for goods is received the clerk enters the product's catalogue number into the computer and is then presented with a monitor display giving details of product availability and the current price. The customer's name and address and their reference number (if they are a previous customer) can also be entered into the computer to check their credit rating. If the goods are available and the customer agrees the current price an order is prepared and is sent to both a printer at the main office for an account to be printed out and to the warehouse. At the warehouse staff can use the data terminal to check orders, to update stock levels as goods are sent to customers, and to print out an invoice that will be sent to the customer along with the goods.

Sometimes more than one computer may be integrated within the data network of a particular organization and in such cases there will also be a need for two computers to be able to communicate directly with one another. The microcomputers used in many offices may be linked together, and also given access to common expensive peripheral equipment, such as printers, by mean of a *local area network* or LAN.

The vast majority of data terminals employ the *International Alphabet 5* code (IA5); this is a 7-bit code that is commonly known as the *ASCII* code (short for American Standard Code for Information Interchange) (see Appendix A). IBM terminals use the EBDIC (Extended Binary Coded Decimal Interchange) 8-bit code.

The data fed into a computer is in binary form with the binary

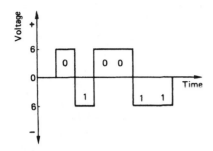

Fig. 1.14 Data waveform

number 0 being represented by a positive voltage and binary 1 by a negative voltage. The actual voltages used are specified by the EIA (Electronic Industries Association) standards given on p. 121 but ±6 V is commonly employed as shown by Fig. 1.14. Each binary digit, 0 or 1, is known as a *bit*. The bit rate is the rate at which information can be transmitted and this is the number of bits transmitted per second. The data waveform of Fig. 1.14 consists of a d.c. component (equal to the average value of the waveform), a fundamental frequency, and a number of harmonics of the fundamental frequency. The fundamental frequency is at its maximum value when the data signal consists of alternate 1s and 0s; this maximum frequency is equal to one-half of the bit rate.

The bit rate of a data signal will be numerically equal to the baud speed whenever the bits are individually transmitted to line. This, however, is only the case for bit rates up to 1200 bits/s. At higher bit rates than 1200 bits are transmitted in groups of either two (dibits), or three (tribits), or four (quabits), and then the baud speed on the line will be less than the bit rate.

Range of Frequencies in Communication

Not all the frequencies present in a speech or music waveform are converted into an electrical signal, transmitted over the communication system, and then reproduced as sound at the distant end. There are two reasons for this: (*a*) for economic reasons the transducers used to convert speech and music signals into electrical form have a limited bandwidth; (*b*) a number of circuits are often conveyed over a single telecommunication system and this practice causes a further limitation of bandwidth. It is therefore desirable to have some idea of the effect on the ear as it responds to a sound waveform, whose frequency components have amplitude relationships differing from those existing in the original sound.

By international agreement the audio frequency band for a 'commercial quality' speech circuit routed over a multi-channel system is restricted to 300−3400 Hz; while for a circuit operated over a high-frequency radio link the bandwidth is only 250−3000 Hz. This means that both the lower and upper frequencies contained in the average speech waveform are not transmitted. To the ear, the pitch of a complex, repetitive, sound waveform is the pitch corresponding to the frequency difference between the harmonics contained in the waveform, i.e. the pitch is that of the fundamental frequency. Hence, even though the fundamental frequency itself may have been suppressed, the pitch of the sound heard by the listener is the same as the pitch of the original sound. However, much of the power contained in the original sound is lost. Suppression of all the frequencies above 3400 Hz reduces the quality of the sound but it does not affect its intelligibility. Since the function of a telephone system is to transmit intelligible speech, the loss of quality can be tolerated; sufficient quality remains to allow a speaker's voice to be recognized.

In the transmission of music the enjoyment of the listener must also be considered. The enjoyment of a person listening to music over a communication system may be considerably impaired if too many of the higher harmonics have been suppressed, since the loss of these harmonics could well result in it becoming difficult, if not impossible, to distinguish between the various kinds of musical instruments. For music circuits routed over line communication systems, a wider bandwidth must be allowed than is allocated to commercial speech circuits. A typical bandwidth in practice is 30 to 10 000 Hz, which makes excellent quality reproduction possible. Land-line music circuits of this kind are used to connect together two BBC studios or a studio and a transmitting station.

When music is broadcast in the long and medium wavebands a bandwidth as wide as 30—10 000 Hz cannot be achieved because the wavebands are shared by so many different broadcasting stations. By international agreement medium waveband broadcasting stations in Europe are spaced approximately 9000 Hz apart in the frequency spectrum, and this means that to make it possible for any particular station to be selected by a radio receiver, without undue interference from adjacent (in frequency) stations, the output sound bandwidth of the receiver cannot be much greater than 4500 Hz. Thus the effective bandwidth of a medium wave broadcast transmission, be it music or speech, is of the order of 50—4500 Hz. Sound broadcast stations in the high-frequency band are allowed an r.f. bandwidth of 10 kHz. This provides an audio bandwidth of 50—5000 Hz. For various reasons the same selectivity is not demanded of VHF sound broadcast receivers or of UHF television receivers, and consequently the audio bandwidths handled by these receivers are somewhat greater. Frequency-modulated sound broadcast transmissions in the 88—108 MHz band provide audio signals up to 15 kHz which allows reasonably good quality reception of musical programmes with a receiver of adequate performance. The video bandwidth allocated to a UHF television signal is limited to 5.5 MHz. This practice leads to some loss of the horizontal definition of the received picture. The bandwidth of the audio signal is 20 kHz.

Exercises

1.1 List the factors that are considered in the selection of the bandwidth to be provided by an audio communication link.

1.2 Briefly discuss the reasons why a circuit provided for the transmission of music must have a wider bandwidth than a speech circuit.

1.3 Draw two curves to show how the thresholds of audibility and feeling vary for the average human ear. Explain the significance of the two points at which the two curves coincide with one another.

1.4 Figure 1.15 shows the basic circuit of a bi-directional telephone link. Explain how it works. What is the main disadvantage of this simple circuit?

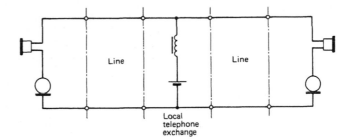

Fig. 1.15

1.5 Which of the following factors are contributory to the bandwidth occupied by a television picture signal: (a) the number of lines per picture; (b) the number of frames per second; or (c) the duration and number of synchronization pulses per second? State the values used in the UK system. Why does a colour TV signal occupy the same bandwidth as a monochrome signal?

1.6 A data signal consists of marks and spaces, each of which has a time duration of 104.17 μs. Calculate (a) the bit rate, (b) the baud speed of the signal.

1.7 Calculate the baud speeds of the two waveforms given in Fig. 1.16.

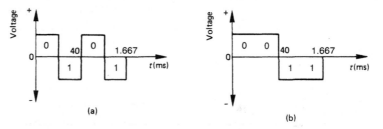

Fig. 1.16

1.8 A data circuit is operated at a speed of 2400 bauds. Calculate the duration of a transmitted bit. Also calculate the maximum fundamental frequency of the waveform.

1.9 Explain why the full range of frequencies that the human voice is capable of generating and the human ear is able to hear is not transmitted over a communication system.

1.10 When music signals are transmitted over land lines the bandwidth provided is of the order of 50–10 000 Hz. Yet the bandwidth provided by medium-wave broadcast stations is only about 4500 Hz. Explain why this is so.

1.11 A data waveform has a bit rate of 14 400 bits/s. Calculate: (a) the baud speed; (b) the duration of a bit.

1.12 Draw a data waveform that has a bit rate of 4800 bits/s. Assume the ASCII code and the use of start and stop bits, and odd parity.

2 Frequency, Wavelength and Velocity

Phase Velocity, Frequency and Wavelength

The propagation of a voltage or current wave along a telephone line, or of a radio wave through the atmosphere, is not an instantaneous process but, instead, occupies a definite interval of time. There is thus a time lag between the application of an a.c. voltage at one end of a line and the detection of the resulting voltage change at a distant point along the line. This means that there will be a *phase difference* between the a.c. voltages existing at two points along a line at a given instant, since the instantaneous voltage at each point is continuously changing. For example, suppose that for a given line the voltage at a point x miles from the sending end lags behind the sending end voltage by 90°. Then at a distance of $2x$ miles from the sending end the voltage will be lagging by $2 \times 90°$ or 180° behind the sending-end voltage. The voltage $4x$ miles from the sending end will be in phase with the sending-end voltage, i.e. a complete cycle of instantaneous values has been completed. The distance $4x$ is known as *one wavelength*, symbol λ (see Fig. 2.1). The wavelength λ is the distance, in metres, between two similar points, in the propagating waveform, e.g. the distance between two successive positive peaks as shown.

Suppose a sinusoidal voltage of frequency f is applied to the input terminals of a transmission line. The periodic time of this waveform is $T = 1/f$ seconds. In a time T seconds the voltage wave will travel along the line a distance equal to T times the velocity v with which it is propagated. During this time the voltage at the sending end of the line will have described one complete cycle of variation with time. While this happens the wave travels a distance that is the wavelength λ of the signal. Therefore

$$\lambda = vT = v/f$$

or

$$v = \lambda f \qquad (2.1)$$

A radio wave travelling through the atmosphere has its velocity of propagation equal to the velocity of light c, which is usually taken as 3×10^8 m/s. The velocity of propagation of a voltage, or a current, wave along a transmission line is less than 3×10^8 m/s, the

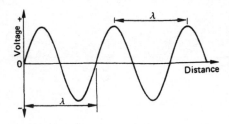

Fig. 2.1 Wavelength of a sinusoidal signal

the amount of reduction depending upon the inductance and capacitance of the line.

Example 2.1

A 30 MHz radio wave is propagated through the atmosphere. Calculate its wavelength.

Solution
From equation (2.1)

$$\lambda = (3 \times 10^8)/(30 \times 10^6) = 10 \text{ m} \quad (Ans.)$$

Example 2.2

The BBC broadcasting station Radio 5 is broadcast on a wavelength of 330 m. What is its frequency?

Solution

$$f = (3 \times 10^8)/330 = 909.1 \text{ kHz} \quad (Ans.)$$

Example 2.3

A 5 MHz radio transmitter is connected to an aerial by a length of transmission line. If the wavelength of a signal on the transmission line is 0.8 times the wavelength of a signal in the atmosphere, calculate the velocity with which a signal is propagated along the cable.

Solution
For the wave in the atmosphere,

$$\lambda_{\text{air}} = c/f = (3 \times 10^8)/(5 \times 10^6) = 60 \text{ m}$$

Hence, the wavelength in the cable is

$$\lambda_{\text{cab}} = 0.8 \times 60 = 48 \text{ m}$$

and therefore the velocity of propagation in the transmission line is

$$v_{\text{p}} = \lambda_{\text{cab}} f = 48 \times 5 \times 10^6 = 2.40 \times 10^8 \text{ m/s} \quad (Ans.)$$

Example 2.4

If the maximum frequency tolerance for fixed-service radio stations operating in the frequency band 4.0 to 27.5 MHz is ± 30 parts in 10^6 ($\pm 0.003\%$) and for radio stations operating in the frequency band 40.0 to 70.0 MHz is ± 10 parts in 10^6 ($\pm 0.001\%$), what frequency variations in hertz do these tolerances correspond to for

- (*a*) a radio station on 12.3 metres,
- (*b*) a radio station on 5.5 metres?
- (*c*) At what wavelength does a station in band (a) radio, having a variation of ± 300 Hz, become outside the permitted tolerance?

Solution

$f = v/\lambda$, where f = frequency in hertz, v = velocity in m/s, and λ = wavelength in metres.

(a) $\lambda = 12.3$ m:

$$f = (3 \times 10^8)/12.3 = 24.39 \text{ MHz}$$

The maximum frequency tolerance in this band is ± 30 parts in 10^6.

Frequency tolerance = $\pm 30/10^6 \times 24.39 \times 10^6 =$
± 732 Hz   (*Ans.*)

(b) $\lambda = 5.5$ m:

$$f = (3 \times 10^8)/5.5 = 54.55 \text{ MHz}$$

The maximum frequency tolerance in this band is ± 10 parts in 10^6.

Frequency tolerance = $\pm 10/10^6 \times 54.55 \times 10^6 = 545.5$ Hz   (*Ans.*)

(c) Frequency variation allowed = frequency $\times$ frequency tolerance

Therefore

$$\pm 300 = \text{frequency} \times \pm 30/10^6$$

$$\text{Frequency} = (300 \times 10^6)/30 = 10 \text{ MHz}$$

Therefore

$$\lambda = (3 \times 10^8)/(10 \times 10^6) = 30 \text{ m}   (\textit{Ans.})$$

Exercises

2.1   Calculate the velocity of a signal of frequency 50 kHz and wavelength 5200 m.

2.2   A radio-frequency transmission line is $\lambda/4$ long at 300 MHz. Calculate the length of the line, in metres.

2.3   Calculate the wavelength of a 60 kHz signal propagating at the velocity of light.

2.4   A radio station has a maximum frequency tolerance of ± 10 parts in 10^6 and operates at 100 MHz. Calculate the maximum frequency change in hertz.

2.5   A radio station operating at a frequency of 90 MHz has a maximum allowable carrier frequency variation of ± 450 Hz. Calculate the frequency tolerance in parts per million.

2.6   Two radio stations operate at frequencies that are 15 kHz apart. Express the wavelength separation in metres. The lower frequency station operates at 1 MHz.

3 Modulation

The bandwidth required for the transmission of commercial-quality speech is 300–3400 Hz and as long as a physical pair of wires, or two pairs of wires, are provided to connect the two parties to a conversation no problem exists. Telephone cable is, however, very expensive and, together with the associated duct work and manholes, comprises the major part of the cost of linking two points. It is desirable, therefore, to be able to transmit more than one conversation over a given link and thus economize in telephone cable. If a number of telephone conversations were merely transmitted into one end of a line it would not be possible to separate them at the distant end of the line since each conversation would be occupying the same part of the frequency spectrum. How then can many different conversations be transmitted over a single circuit and yet be separable at the distant end? Two main methods exist: in one, known as *frequency-division multiplexing* (FDM), each conversation is shifted, or translated, to a different part of the frequency spectrum. In the other, known as *time-division multiplexing* (TDM), the conversations each occupy the same frequency band but are applied in time sequence to the common line. Both types of multiplexing improve the efficiency with which line plant is used but TDM is less affected by noise on the line.

Frequency-division Multiplex

To illustrate the principle of an FDM system, consider the simple case where it is required to transmit three telephone channels, of bandwidth 300 to 3400 Hz, over a common line. The first of these channels can be transmitted directly over the common line, or bearer circuit, and it will then occupy the band 300 to 3400 Hz. The second and third channels cannot also be transmitted directly over the line because they would be inseparable from the first channel and from each other. Suppose, therefore, that instead these two channels are each passed into a circuit which frequency translates, or shifts, them to the frequency bands 4300 to 7400 Hz and 8300 to 11 400 Hz respectively, before transmission to line. The three channels can now all be transmitted over the bearer circuit but since there is a frequency gap of 900 Hz between them no inter-channel interference will occur. At the receiving end of the bearer circuit, filters separate the three channels and further circuits restore the second and third channels to their original frequency bands. The bandwidth provided by the bearer circuit must be 300–11 400 Hz.

The frequency translation of a channel to a position higher in the frequency spectrum is known as *modulation* and the circuit which achieves it as a modulator. The particular part of the frequency spectrum to which the channel is shifted is determined by the frequency of the sinusoidal *carrier wave* which is modulated. Modulation may be defined as the process by which one of the characteristics of a carrier wave is modified in accordance with the characteristics of a modulating signal. The restoration of a channel to its original position in the frequency spectrum is known as demodulation, the circuit used is known as a demodulator.

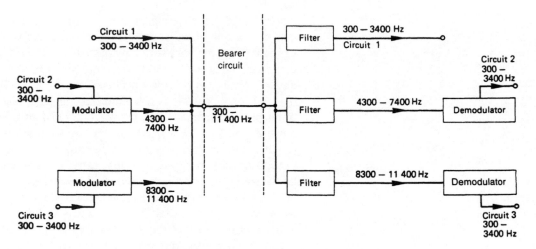

Fig. 3.1 A simple 3-channel FDM system

A block schematic diagram of the equipment required for one direction of transmission in the three-channel FDM system just described is shown in Fig. 3.1. Since the equipment is unidirectional, it needs to be duplicated for transmission in both directions. Furthermore, the line may be a telephone cable or may be a VHF, UHF or microwave radio link or a communication satellite system. FDM multi-channel telephone systems are no longer employed in the UK trunk network.

Frequency translation is also employed for radio and television broadcasting. The broadcasts are transmitted by the BBC and by the Independent Broadcasting Authority (IBA) and received in the home by means of an aerial, but no type of aerial is able to operate at audio frequencies. It is therefore necessary to shift each programme originally produced in the audio-frequency band to a point higher in the frequency spectrum where aerials can operate with reasonable efficiency. Since there is usually a large number of broadcasting stations within a given geographical area, it is necessary to arrange that each station, as far as possible, occupies a different part of the usable frequency spectrum. Hence the programmes radiated by different broadcasting stations are frequency translated to their own, internationally agreed, fixed frequency bands. For example, consider the medium waveband stations BBC Radios 1 and 5. Radio 1 is broadcast on frequencies of 1053 and 1089 kHz and Radio 5 on 909 kHz.

The term *baseband* is often employed to describe the band of frequencies that must be transmited by a system for an adequate signal to appear at the receiving terminal. The baseband signal will usually not contain all the frequencies generated by the source; for example, the speech baseband signal is 300−3400 Hz.

Time-division Multiplex

With time-division multiplex (TDM) a number of different channels can be transmitted over a common bearer circuit by allocating the bearer circuit to each channel in turn for a given period of time, i.e. at any particular instant only one channel is connected to the bearer circuit. The principle of a TDM system is illustrated by Fig. 3.2 which shows the basic arrangement of a two-channel TDM system.

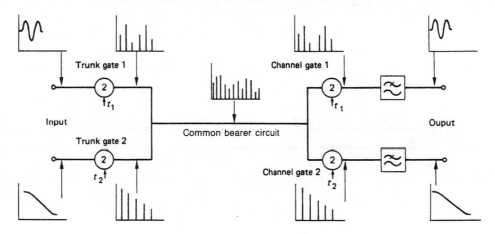

Fig. 3.2 A simple 2-channel TDM system. t_1 = a series of pulses occurring at fixed intervals. t_2 = a series of pulses occurring at the same periodicity as t_1 but commencing later by an amount equal to half the time interval (From *B.T.E. Journal*)

The two channels which are to share the common circuit are each connected to it via a *channel gate*. The channel gates are electronic switches which only permit the signal present on a channel to pass when opened by the application of a controlling pulse. Hence, if the controlling pulse is applied to gate 1 at time t_1 and not to gate 2, gate 1 will open for a time equal to the duration of the pulse but gate 2 will remain closed. During this time, therefore, a sample of the instantaneous signal waveform on channel 1 will be transmitted to line. At the end of the pulse, both gates are closed and no signal is transmitted to line. If now the controlling pulse is applied to gate 2 at a later time t_2, gate 2 will open and a sample of the signal waveform on channel 2 will be transmitted over the bearer circuit. Thus if the pulses applied to control the opening and shutting of gates 1 and 2 are repeated at regular intervals, a series of samples of the signal waveforms existing on the two channels will be transmitted.

At the receiving end of the system gates 1 and 2 are opened, by the application of control pulses, at those instants when the incoming waveform samples appropriate to their channel are being received. This requirement demands accurate *synchronization* between the controlling pulses applied to the gates 1, and also between the controlling pulses applied to the gates 2. If the time taken for signals to travel over the bearer circuit was zero then the system would require controlling pulses as shown in the figure, but since, in practice, the transmission time is not zero, the controlling pulses applied at the receiving end of the system must occur slightly later than the corresponding controlling pulses at the sending end. If the pulse synchronization is correct the waveform samples are directed to the correct channels at the receiving end. The received samples must then be converted back to the original waveform, i.e. demodulated. Provided the sampling rate is at least equal to twice the highest frequency contained in the original signal waveform, the pulse waveform can be demodulated merely by passing the samples through a low-pass filter that passes all frequencies lower than the sampling frequency.

Types of Modulation

For a signal to be frequency translated to another part of the frequency spectrum it is necessary for the signal to vary one of the characteristics of a sinusoidal wave — usually known as the *carrier wave* — whose frequency occupies the required part of the spectrum. The process by which one of the characteristics of a carrier wave is modified in accordance with the characteristics of a modulating signal is known as modulation.

The general expression for a sinusoidal carrier wave is

$$v = V \sin (\omega t + \theta) \tag{3.1}$$

where v = instantaneous voltage of the wave,
V = peak value or amplitude of the wave,
ω = angular velocity of the wave in radians/second; ω is related to the frequency of the wave by the expression $\omega = 2\pi f$ where f is the frequency in hertz,
θ = the phase of the wave at the instant when $t = 0$.

For modulation is it necessary to cause one of the characteristics of the wave to be varied in accordance with the waveform of the modulating signal. Three possibilities exist:

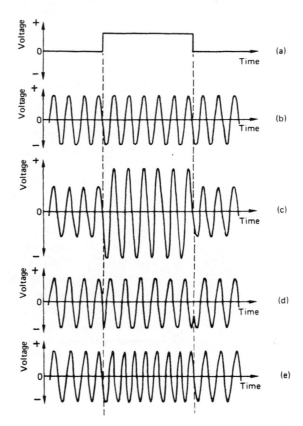

Fig. 3.3 Illustrating the difference between amplitude, frequency, and phase modulation
(a) Modulating signal
(b) Unmodulated carrier wave
(c) Amplitude-modulated wave
(d) Phase-modulated wave
(e) Frequency-modulated wave

(a) the amplitude V of the wave may be varied to give amplitude modulation,

(b) the frequency $\omega/2\pi$ may be modified to give frequency modulation,

(c) the phase θ may be the variable, giving phase modulation.

To illustrate, graphically, the difference between these three types of modulation, Fig. 3.3 shows the waveform produced when the modulating signal is a positive voltage pulse. In the case of the *amplitude-modulated* (AM) wave, Fig. 3.3(c), the frequency of the carrier wave remains constant but its amplitude is suddenly increased when the leading edge of the modulating pulse occurs and then suddenly reduced to its original value when the pulse ends. The *phase-modulated* (PM) carrier wave shown in Fig. 3.3(d) has a constant amplitude and frequency, but experiences an abrupt change of phase at the instants corresponding to the beginning and end of the modulating signal pulse. Finally, the *frequency-modulated* (FM) carrier wave (Fig. 3.3(e)) also has a constant amplitude, but the frequency of the wave is abruptly increased at the beginning of the modulating pulse and maintained at the new frequency until the pulse ends, when the frequency is suddenly decreased to its original value.

Amplitude modulation is employed for FDM telephony systems, radio broadcasting in the long, medium and short wavebands, the picture signal of 625-line television broadcasts and for various point-to-point and base-to-mobile radio-telephony systems.

Frequency modulation is used for data links, VHF radio broadcasting, the sound signal of 625-line television broadcasting, as well as for some base-to-mobile radio-telephony systems. Phase modulation is employed in data transmission links and as a stage in the production of a frequency-modulated signal. Frequency modulation may have an advantage over amplitude modulation in that its performance in the presence of interfering signals and noise can be much better, but it then suffers from the disadvantage of requiring a larger bandwidth. Because of the large bandwidth required for a frequency-modulated system the use of frequency modulation for broadcasting and for radio-telephony is restricted to VHF and UHF where the bandwidth required can be more readily provided.

Systems using the principle of time-division multiplex are generally based on the sampling of the amplitude of the information signal at regular intervals, and the subsequent transmission of a pulse signal to represent each sample. Four main types of pulse modulation exist, namely pulse-amplitude modulation (PAM), pulse-duration modulation (PDM), pulse-position modulation (PPM) and pulse-code modulation (PCM). The principle of each of the first three types of pulse modulation is shown in Fig. 3.4.

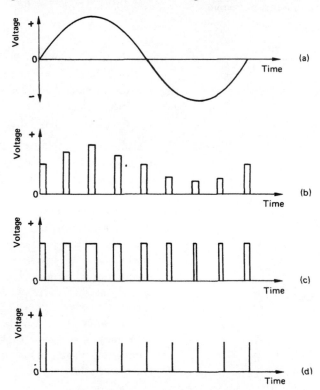

Fig. 3.4 Illustrating the difference between PAM, PDM and PPM
(a) Modulating signal
(b) Pulse-amplitude modulation
(c) Pulse-duration modulation
(d) Pulse-position modulation

With PAM, pulses of equal width and common spacing in time are employed and have their amplitudes varied in proportion to the instantaneous amplitude of the modulating signal. PDM employs pulses of constant amplitude and common spacing between their leading edges, but whose width, or duration, is varied in accordance with the waveform of the modulating signal. In the figure the leading edges of the pulses occur at constant time intervals and the trailing edges are modulated in time. Finally PPM employs constant amplitude and width pulses whose position (in time) is dependent upon the instantaneous amplitude of the modulating signal. Each narrow pulse occurs at the same time as the modulated trailing edges of the PDM pulses.

Each of these three methods of pulse modulation suffers from the disadvantage that distortion and noise is cumulative along the system — although the PPM system is much less affected than the other two in this way. The fourth type of pulse modulation, i.e. pulse code modulation (PCM), is a more sophisticated method which largely overcomes the effects of distortion and noise, but at the expense of added complexity and a wider bandwidth. PCM is considered in Chapter 12.

Amplitude Modulation

If a carrier wave is amplitude modulated, the amplitude of the carrier is caused to vary in accordance with the instantaneous value of the modulating signal. For example, consider the simplest case when the

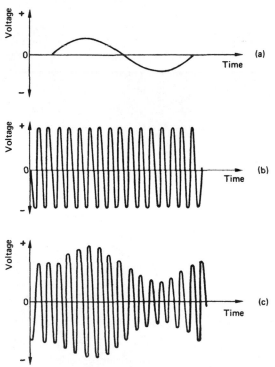

Fig. 3.5 A carrier wave modulated by a sinusoidal signal
(a) Modulating signal
(b) Unmodulated carrier wave
(c) Amplitude-modulated carrier wave

modulating signal is itself sinusoidal. The amplitude of the modulated wave must then vary in a sinusoidal manner as shown in Fig. 3.5. The difference in frequency between the carrier wave and the modulating signal will normally be much higher than is indicated by the diagram. In practice, the frequency difference would at least be of some thousands of hertz.

The amplitude of the modulated carrier wave can be clearly seen to vary between a maximum value, which is greater than the amplitude of the unmodulated wave, and a minimum value, which is less than the amplitude of the unmodulated wave. The outline of the modulated carrier wave is known as the modulation *envelope*.

To convey information, a signal containing at least two components at different frequencies is required and hence, in practice, a sinusoidal modulating signal is rarely used. However, whatever the waveform of the modulating signal the same principle still holds good: the envelope of the modulated wave must be the same as the waveform of the modulating signal. For example, Fig. 3.6 shows the envelopes of a carrier wave modulated by (*a*) a signal consisting of components at a fundamental frequency and its third harmonic, both components being in phase at time $t = 0$, and (*b*) note C played on a cello organ pipe.

The Frequencies in an Amplitude-modulated Wave

When a sinusoidal carrier wave is amplitude modulated, each frequency component in the modulating signal gives rise to two frequencies in the modulated wave, one below the carrier frequency and one above. When, for example, the modulating signal is a sinusoidal wave of frequency f_m the modulated carrier wave contains three frequencies:

(*a*) the carrier frequency f_c;
(*b*) the lower sidefrequency $(f_c - f_m)$; and
(*c*) the upper sidefrequency $(f_c + f_m)$.

The two new frequencies are the upper and lower *sidefrequencies* $(f_c + f_m)$ and $(f_c - f_m)$ respectively, and these are equally spaced either side of the carrier frequency by an amount equal to the modulating signal frequency f_m. The frequency f_m of the modulating signal itself is *not* present.

The bandwidth required to transmit an amplitude-modulated carrier wave is equal to the difference between the highest frequency to be transmitted and the lowest frequency. In the case of sinusoidal modulation and with $f_c > f_m$, the bandwidth required is given by

$$B = (f_c + f_m) - (f_c - f_m) = 2f_m \tag{3.2}$$

i.e. the required bandwidth is equal to twice the frequency of the modulating signal.

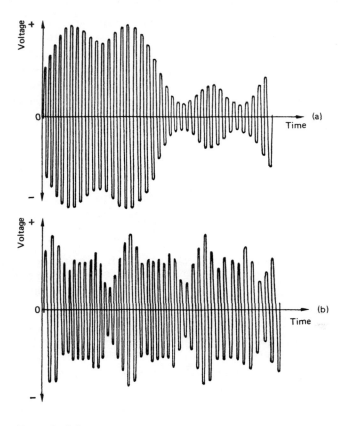

Fig. 3.6 Waveforms of a carrier wave modulated by (a) a fundamental plus third harmonic (b) a cello organ pipe playing C

Example 3.1

A 100 kHz carrier is amplitude modulated by a sinusoidal tone of frequency 3000 Hz. Determine the frequencies contained in the amplitude-modulated wave and the bandwidth required for its transmission.

Solution

The frequencies contained in the modulated wave are

(a) The carrier frequency $f_c = 100\,000$ Hz   (*Ans.*)
(b) The lower sidefrequency $(f_c - f_m) = 100\,000 - 3000$
$$= 97\,000 \text{ Hz}   (Ans.)$$
(c) The upper sidefrequency $(f_c + f_m) = 100\,000 + 3000$
$$= 103\,000 \text{ Hz}   (Ans.)$$
   The bandwidth required $= 103\,000 - 97\,000 = 6000$ Hz   (*Ans.*)

If the modulating signal is non-sinusoidal it will contain components at a number of different frequencies; suppose that the highest frequency contained in the modulating signal is f_2 and the lowest frequency is f_1. Then the frequency f_2 will produce an upper sidefrequency component $(f_c + f_2)$ and a lower sidefrequency component $(f_c - f_2)$, while frequency f_1 will produce upper and lower sidefrequencies $(f_c \pm f_1)$. Thus the modulating signal will produce a number of lower sidefrequency components lying in the range $(f_c - f_2)$ to $(f_c - f_1)$ and a number of upper sidefrequency components in the range $(f_c + f_1)$ to $(f_c + f_2)$. The band of frequencies below the carrier frequency, i.e. $(f_c - f_2)$ to $(f_c - f_1)$, is known as the *lower sideband*, while the band of frequencies above the carrier frequency is known as the *upper sideband*. When the carrier frequency is higher than the modulating signal frequency the sidebands are symmetrically situated, with respect to frequency, on either side of the carrier frequency. The lower sideband is said to be *inverted* because the highest frequency in it $(f_c - f_1)$ corresponds to the lowest frequency f_1 in the modulating signal and vice versa. Similarly, the upper sideband is said to be *erect* because the lowest frequency in it $(f_c + f_1)$ corresponds to the lowest frequency f_1 in the modulating signal.

Example 3.2

A 108 kHz carrier wave is amplitude modulated by a band of frequencies, 300–3400 Hz. What frequencies are contained in the upper and lower sidebands of the amplitude-modulated wave and what is the bandwidth required to transmit the wave?

Solution
The lower sideband will contain frequencies in the band

$\quad$ 108 000 − 3400 Hz to 108 000 − 300 Hz,
$\quad$ or 104 600 to 107 700 Hz $\quad$ (*Ans.*)

The upper sideband will contain frequencies between

$\quad$ 108 000 + 300 Hz and 108 000 + 3400 Hz,
$\quad$ or 108 300 and 111 400 Hz $\quad$ (*Ans.*)

The required bandwidth B is equal to the maximum frequency contained in the modulated wave minus the minimum frequency:

$\quad B = 111\,400 - 104\,600 = 6800\,\text{Hz} \quad$ (*Ans.*)

Note that the bandwidth is equal to twice the highest frequency contained in the modulating signal.

It is possible to confirm graphically the presence of a number of frequencies in an amplitude-modulated wave. Consider, for example, the case of a 10 000 Hz carrier wave and a 2000 Hz modulating signal. The upper sidefrequency will be 12 000 Hz and the lower sidefrequency will be 8000 Hz. The waveforms of the carrier, the upper sidefrequency and the lower sidefrequency components are shown in Fig. 3.7(*a*), (*b*) and (*c*) respectively. The carrier component can be seen to complete 10 cycles, the upper sidefrequency component 12 cycles and the lower sidefrequency component 8 cycles in the time of one millisecond. To obtain the waveform of the amplitude-modulated wave, which is composed of these three components, it is necessary to sum the instantaneous values of the three waves. It can be seen that the envelope of the modulated wave shows two

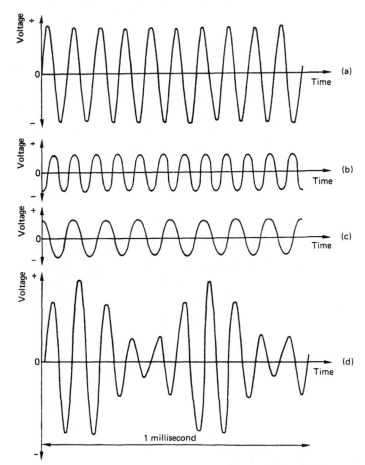

Fig. 3.7 Showing the formation of an amplitude-modulated wave by adding the components at the carrier and upper and lower sidefrequencies
(*a*) Carrier wave
(*b*) Upper sidefrequency
(*c*) Lower sidefrequency
(*d*) Amplitude-modulated carrier wave

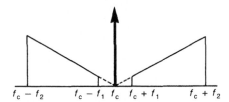

Fig. 3.8 Frequency spectrum of an amplitude-modulated wave

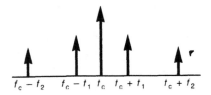

Fig. 3.9 A method of representing the sidebands of amplitude modulation

complete variations in the time of one millisecond, i.e. it varies at the modulating frequency of 2000 Hz.

There are two ways in which the frequency spectrum occupied by a modulated carrier wave may be shown. Each component may be represented by an arrow drawn perpendicular to the frequency axis as shown in Fig. 3.8, where f_c, f_1 and f_2 have the same meanings as before. The lengths of the arrows are made proportional to the amplitude of the component they each represent. If many of the frequencies between f_1 and f_2 were present in the modulating signal, a large number of arrows would be required and the diagram would become impracticable. Alternatively, the sidebands produced by a complex modulating signal can be represented by truncated triangles, in which the vertical ordinates are made proportional to the modulating frequency and no account is taken of amplitude — see Fig. 3.9. This method of representing sidebands gives an immediate indication of which sideband is erect and which is inverted. This is useful when considering systems employing more than one stage of modulation when the inverted sideband is not necessarily the lower sideband.

Modulation Depth

The envelope of an amplitude-modulated carrier wave varies in accordance with the waveform of the modulating signal and hence there must be a relationship between the maximum and minimum values of the modulated wave and the amplitude of the modulating signal. This relationship is expressed in terms of the modulation factor of the modulated wave.

The *modulation factor m* of an amplitude-modulated wave is defined by the expression

$$m = \frac{\text{maximum amplitude} - \text{minimum amplitude}}{\text{maximum amplitude} + \text{minimum amplitude}} \quad (3.3)$$

When expressed as a percentage m is known as the modulation depth, or the depth of modulation or the percentage modulation.

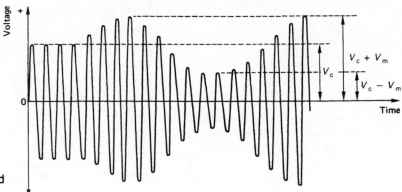

Fig. 3.10 Characteristics of an amplitude-modulated wave required to calculate modulation depth

Figure 3.10 shows a sinusoidally modulated wave. Since the envelope of the modulated carrier wave must vary in accordance with the modulating signal, its maximum amplitude must be equal to the amplitude of the carrier wave plus the amplitude of the modulating signal, i.e. $(V_c + V_m)$. Similarly, the minimum amplitude of the modulated wave must be equal to $(V_c - V_m)$, where V_c is the amplitude of the carrier wave and V_m is the amplitude of the modulating signal.

For *sinusoidal modulation*, therefore, the modulation factor m is equal to

$$m = \frac{(V_c + V_m) - (V_c - V_m)}{(V_c + V_m) + (V_c - V_m)} = \frac{V_m}{V_c} \quad (3.4)$$

The modulation factor is equal to the ratio of the amplitude of the modulating signal to the amplitude of the carrier wave.

Also, for a sinusoidally modulated carrier wave, the two sidefrequencies have the same amplitude of $m/2$ times the amplitude of the carrier wave.

Example 3.3

Draw the waveform of an amplitude-modulated carrier wave that is sinusoidally modulated to a depth of 25%.

Solution

For a modulation depth of 25%, $m = 0.25$, and since the modulating signal is sinusoidal, $m = V_m/V_c$ or $V_m = 0.25 V_c$. Thus

the maximum amplitude of the modulated wave is $V_c + 0.25 V_c = 1.25 V_c$

and the minimum amplitude is $V_c - 0.25 V_c = 0.75 V_c$

The required waveform is shown in Fig. 3.11.

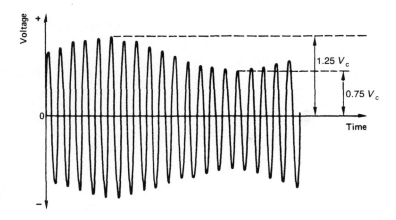

Fig. 3.11 Amplitude-modulated wave of modulation depth 25%

Example 3.4

Determine the depth of modulation of the amplitude-modulated wave shown in Fig. 3.12.

Solution

The maximum voltage of the wave is 10 V and the minimum voltage is 5 V; hence the depth of modulation is

$$(10 - 5)/(10 + 5) \times 100\% = 33.3\% \quad (Ans.)$$

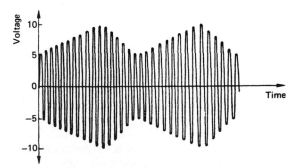

Fig. 3.12 Carrier wave amplitude modulated by a non-sinusoidal waveform

Example 3.5

The envelope of a sinusoidally modulated carrier wave varies between a maximum value of 8 V and a minimum value of 2 V. Find (*a*) the amplitude of the carrier frequency component, (*b*) the amplitude of the modulating signal and (*c*) the amplitudes of the two sidefrequencies.

Solution

(*a*) Maximum value $V_c + V_m = 8$ V (3.5)

Minimum value $V_c - V_m = 2$ V (3.6)

Adding equations (3.5) and (3.6)

$2V_c = 10$ or $V_c = 5$ V (*Ans.*)

(*b*) Subtracting equation (3.6) from equation (3.5)

$2V_m = 6$ or $V_m = 3$ V (*Ans.*)

(*c*) The amplitude of each of the two sidefrequencies is equal to

$mV_c/2$ i.e. $V_m/V_c \times V_c/2$ or $V_m/2$

Amplitude of sidefrequencies = 3/2 or 1.5 V (*Ans.*)

When the depth of modulation is 100% the modulation envelope varies between a maximum of $2V_c$ and a minimum of 0. If the depth of modulation is increased beyond this value the modulation envelope becomes distorted as shown in Fig. 3.13. The carrier is then said to be *over-modulated*.

Since the envelope of the modulated wave is no longer a replica

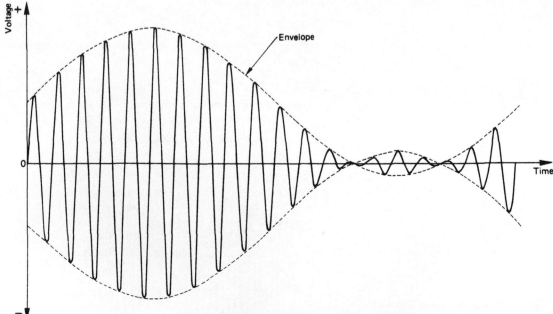

Fig. 3.13 Amplitude-modulated wave with a modulation depth in excess of 100%

of the modulating signal waveform, considerable distortion has taken place. This, of course, means that modulation depths greater than 100% are never employed in practice.

Power of an Amplitude-modulated Wave

When an amplitude-modulated voltage wave is applied across a resistance R, each component frequency of the wave will dissipate power. The total power dissipated by the wave is the sum of the powers dissipated by the individual components. In the case of a sinusoidally modulated carrier wave, three components exist; namely, the carrier frequency, the upper sidefrequency, and lower sidefrequency.

The power P_c developed by each of the sidefrequency components is

$$P_c = (V_c/\sqrt{2})^2/R = V_c^2/2R \text{ watts}$$

and the power developed by each of the sidefrequency components is

$$\begin{aligned}
P_{LSF} = P_{USF} &= (V_m/2\sqrt{2})^2/R \\
&= (mV_c/2\sqrt{2})^2/R \\
&= (m^2V_c^2)/8R \text{ watts}
\end{aligned}$$

so that the total power P_t is

$$P_t = V_c^2/2R + m^2V_c^2/8R + m^2V_c^2/8R$$

or

$$P_t = (V_c^2/2R)(1 + m^2/2) \text{ watts} \tag{3.7}$$

Example 3.6

The total power dissipated by an amplitude-modulated wave is 1575 W. Calculate the power in the sidefrequencies if the modulation depth is (a) 50% and (b) 100%.

Solution
From equation (3.7), where P_c is the carrier power,

$$1575 = P_c(1 + m^2/2)$$

(a) When $m = 0.5$

$$1575 = P_c(1 + 0.25/2)$$

$$P_c = 1575/1.125 = 1400 \text{ W}$$

Therefore $P_{SF} = 1400 \times 0.125 = 175 \text{ W}$ (*Ans.*)

(b) When $m = 1$

$$1575 = P_c(1 + 0.5)$$

$$P_c = 1575/1.5 = 1050 \text{ W}$$

Therefore $P_{SF} = 1050 \times 0.5 = 525 \text{ W}$ (*Ans.*)

It is clear from this example that the power contained in the two sidefrequencies is only a small fraction of the total power, rising to a maximum, when $m = 1$, equal to one-third of the total power. Since only the sidefrequencies carry information, amplitude modulation is not a very efficient system when considered on a power basis.

Single-sideband Operation

It is clear that an amplitude-modulated wave (DSBAM) contains the intelligence represented by the modulating signal in both the upper sideband and the lower sideband. It is therefore unnecessary to transmit both sidebands. Furthermore, the carrier component is of constant amplitude and frequency and so it does not carry any information. It is possible to suppress both the carrier and one sideband in the transmitting equipment, and to transmit just the other sideband without any loss of information. This method of operation is known as *single-sideband suppressed-carrier* (SSBSC) working. The frequency spectrum diagram of an SSB signal is shown in Fig. 3.14. SSB operation of a communication system offers the following advantages over double-sideband (DSB) working.

(*a*) The bandwidth required for SSB transmission is only half as great as that required for DSB transmission carrying the same information. This allows more channels to be operated within the frequency spectrum provided by the transmission medium.

(*b*) The signal-to-noise ratio* at the receiving end of an SSB system is greater than that of a DSB system. The improvement is 9 dB for a depth of modulation of 100% and even more for modulation depths of less than 100%; some of this improvement is the result of an increase in the ratio sideband power/total power of the transmitted output and the rest is because the necessary bandwidth is reduced by half (noise power is proportional to bandwidth).

(*c*) A DSB transmitter produces a power output (due to the transmitted carrier) at all times, whereas an SSB transmitter does not. A saving in the d.c. power taken from the power supply is thus obtained, with an overall increase in transmitter efficiency.

(*d*) Radio waves are subject to a form of interference known as *selective fading*. When this is prevalent, considerable distortion of a DSB signal may occur, because the carrier component may fade below the level of the sidebands, so that the two sidebands beat together to produce a large number of unwanted frequencies. This cannot occur with an SSB system because the signal is demodulated against a locally generated carrier of constant amplitude.

The disadvantage of SSB working is the need for more complex receiving equipment. The increased complexity is due to the need to reintroduce a carrier at the same frequency as the carrier originally

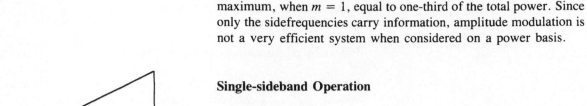

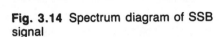

$f_c - f_{m2}$ $f_c - f_{m1}$

Fig. 3.14 Spectrum diagram of SSB signal

* Signal-to-noise ratio is discussed in Chapter 8.

suppressed at the transmitter. Any lack of synchronization between the suppressed and reintroduced carrier frequencies produces a shift in each component frequency of the demodulated signal. To preserve intelligibility the reinserted carrier must be within a few hertz of the original carrier frequency. The extra complexity of operation is the reason why SSB working has been restricted to long distance radio-telephony systems and is not employed for domestic radio broadcasting. Now that complex circuitry is readily available in ICs, SSB sound broadcast receivers are technically feasible, but there are so many conventional broadcast receivers in use that any change to SSB would be politically unacceptable.

Frequency Modulation

If a carrier wave is frequency modulated, its frequency is made to vary in accordance with the instantaneous value of the modulating signal. The amount by which the carrier frequency deviates from its unmodulated value is proportional to the amplitude of the modulating signal, and the number of times per second the carrier deviates is equal to the modulating frequency. Figure 3.15(*b*) shows a frequency-modulated wave and Fig. 3.15(*a*) shows the corresponding modulating

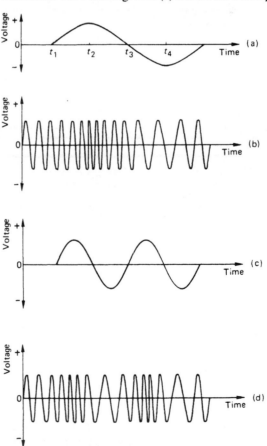

Fig. 3.15 Frequency-modulated waves

signal. Over the time interval 0 to t_1 the modulating signal voltage is zero and so the carrier frequency is unmodulated. From t_1 to t_2 the modulating signal voltage is increasing in the positive direction and the carrier frequency increases to a maximum value at t_2. After time t_2 the modulating voltage falls towards zero, and the carrier frequency reduces in value until at time t_3 it has returned to its unmodulated value. During the following negative half-cycle of the modulating signal voltage, the carrier frequency is reduced below its unmodulated value. The carrier frequency has its minimum value when the modulating signal voltage reaches its peak negative value, i.e. at time t_4.

If the frequency of the modulating signal is increased (Fig. 3.15(c)), the number of times per second that the carrier frequency is varied about its mean, unmodulated, value is increased in proportion. This is shown by Fig. 3.15(d). It should be noticed that the minimum and maximum values of the modulated carrier frequency are the same in both Figs 3.15(b) and (d); this is because the respective modulating signals have equal amplitudes.

Since the amplitude of a carrier wave is not changed when it is frequency modulated, there is no change in the power carried by the wave.

Frequency Deviation

The frequency deviation of a frequency-modulated wave is the amount by which the carrier frequency has been changed from its unmodulated value. Frequency deviation is proportional to the amplitude of the modulating signal voltage. There is no inherent maximum to the frequency deviation that can be achieved (unlike amplitude modulation where 100% modulation is the maximum amplitude deviation possible). Since the bandwidth occupied by a frequency-modulated wave is proportional to frequency deviation, a maximum value to the deviation permitted in a particular system must be arbitrarily chosen. The maximum value of frequency deviation that is allowed to occur in a particular system is known as the *rated system deviation*. Since frequency deviation is proportional to modulating signal amplitude, it follows that a maximum value, which depends on the sensitivity of the modulator, is also determined for the modulating voltage.

Example 3.7

The rated system deviation in the BBC VHF sound broadcast system is 75 kHz. What will be the frequency deviation of the carrier if the modulating signal voltage is (a) 50% and (b) 20% of the maximum value permitted?

Solution
(a) Frequency deviation = 75 kHz × 0.5 = 37.5 kHz   (*Ans.*)
(b) Frequency deviation = 75 kHz × 0.2 = 15 kHz   (*Ans.*)

The term *frequency swing* is sometimes applied to a frequency-modulated wave. It refers to the maximum carrier frequency minus the minimum carrier frequency, i.e. the frequency swing is equal to twice the frequency deviation.

Modulation Index

When a carrier voltage is frequency modulated, its phase is also caused to vary. The peak phase deviation produced is equal to the ratio of the frequency deviation to the modulating frequency and is known as the modulation index m_f of the modulated wave. Thus

$$m_f = \text{(frequency deviation)/(modulating frequency)} \quad (3.8)$$

Example 3.8

A carrier is frequency modulated by a 10 kHz sinusoidal wave whose amplitude is one-half the maximum permitted value. If the rated system deviation is 50 kHz determine (*a*) the frequency deviation and (*b*) the phase deviation, of the carrier that is produced.

Solution
(*a*) Frequency deviation = 50 kHz × 0.5 = 25 kHz   (*Ans.*)
(*b*) $m_f = (25 \times 10^3)/(10 \times 10^3) = 2.5 \text{ rad}$ (*Ans.*)

When both the frequency deviation and the modulating frequency are at their maximum permitted values, the modulation index is then known as the *deviation ratio D*, i.e.

$$D = \frac{\text{maximum frequency deviation}}{\text{maximum modulating frequency}} \quad (3.9)$$

The Frequencies in a Frequency-modulated Wave

When a carrier of frequency f_c is frequency modulated by a sinusoidal wave of frequency f_m, the resultant waveform contains components at a number of different frequencies. The modulated wave contains the following frequency components:

(*a*) the carrier frequency f_c
(*b*) first-order sidefrequencies $f_c \pm f_m$
(*c*) second-order sidefrequencies $f_c \pm 2f_m$
(*d*) third-order sidefrequencies $f_c + 3f_m$

and so on. The number of sidefrequencies present in a particular wave depends upon its modulation index; the larger the value of the modulation index the greater the number of sidefrequencies generated. The amplitudes of the various components, including the carrier itself, vary in a complicated manner as the modulation index is increased.

Any component, again including the carrier, may have zero amplitude at a particular value of modulation index. The carrier is zero for $m_f = 2.405$, 5.32 and 8.654.

The bandwidth required to transmit a frequency-modulated wave is not, as might be expected, twice the frequency deviation of the carrier but, instead, is somewhat greater. The expression generally used to determine the bandwidth occupied by a frequency-modulated wave is given by equation (3.10), i.e.

$$\text{Bandwidth} = 2(f_d + f_m) \tag{3.10}$$

where f_d = peak frequency deviation of carrier, f_m = maximum modulating frequency.

Example 3.9

The BBC VHF frequency-modulated sound transmissions are operated with a rated system deviation of 75 kHz and a maximum modulating frequency of 15 kHz. Determine the required bandwidth. Compare this with the bandwidth required by an amplitude-modulation system that provides the same audio bandwidth.

Solution
From equation (3.10),

$$\text{Bandwidth} = 2(75 \times 10^3 + 15 \times 10^3) = 180\,\text{kHz} \quad (Ans.)$$

If the same audio bandwidth were to be provided by an amplitude-modulated system, the necessary r.f. bandwidth would be $f_c \pm 15$ kHz or 30 kHz. Clearly, in this case, frequency modulation is much more expensive in its use of the available frequency spectrum than is amplitude modulation.

The signal-to-noise ratio at the output of a frequency-modulated system is proportional to the deviation ratio and, since D = (maximum frequency deviation/maximum modulating frequency), also to the bandwidth occupied. This means that the signal-to-noise ratio can always be improved at the cost of an increased bandwidth; this is the reason why the BBC frequency-modulated sound broadcast stations are operated in the VHF band, since here the necessary wide bandwidth is available. With narrow-band frequency modulation (NBFM), the deviation ratio is small and the system offers little, if any, improvement in signal-to-noise ratio over the use of amplitude modulation. NBFM is used for land mobile radio systems.

The Relative Merits of Frequency Modulation and Amplitude Modulation

Frequency modulation offers the following advantages over the use of DSB amplitude modulation.

(a) The signal-to-noise ratio at the output of the f.m. receiver can be greater than that of a DSB amplitude-modulation receiver.
(b) The amplitude, and hence the power, of a frequency-modulated wave is constant, allowing more efficient transmitters to be built.

(c) An f.m. receiver possesses the ability to suppress the weaker of two signals simultaneously received at or near the same frequency. This effect is known as the *capture effect*.

(d) The dynamic range, i.e. the range of modulating signal amplitudes that can be transmitted, is much larger.

The disadvantage of frequency modulation is the (generally) wider bandwidth required.

Amplitude modulation is used for long, medium, and short waveband broadcast transmissions, for the picture signal in television broadcasting, for long-distance radio-telephony, for VHF/UHF base-mobile systems, and for various ship and aeroplane radio services. Frequency modulation is used for VHF sound broadcasting, for the sound signal of UHF television broadcasts, for some base-mobile systems, for some ship/aero services, and for wideband radio-telephony systems.

Exercises

3.1 A 2 kHz sinusoidal waveform is pulse-amplitude modulated at a sampling rate of 20 kHz. Draw the modulated waveform.

3.2 Figure 3.16 shows several different waveforms. Identify each of them.

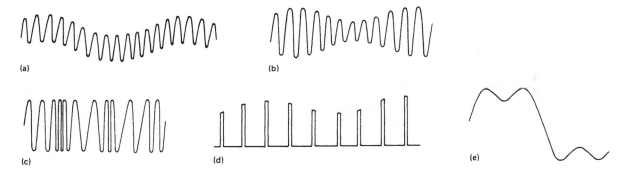

(a)   (b)

(c)   (d)   (e)

Fig. 3.16

3.3 A train of 2 V amplitude, 10 μs-width rectangular pulses are pulse-amplitude-modulated by a 1 V peak sinusoidal signal. Draw the PAM waveform.

3.4 The instantaneous voltage of a sinusoidal carrier wave is given by

$$v = V \sin(\omega t + \theta)$$

Which of these parameters is varied when the wave is:
(a) frequency modulated; (b) amplitude modulated; and (c) phase modulated?

3.5 A 6 V carrier is modulated by a ramp waveform which changes linearly from 0 V to 3 V in 1 ms. Draw the resultant waveform when the carrier is: (a) amplitude modulated; and (b) frequency modulated.

3.6 What kind of waveform is shown in Fig. 3.17? Determine (a) the carrier frequency; (b) the modulating frequency; and (c) the modulation factor.

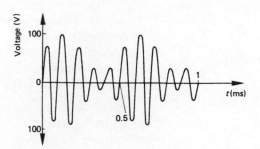

Fig. 3.17

3.7 A 72 kHz carrier wave is amplitude modulated by a 2 kHz wave. Calculate the frequencies contained in the modulated waveform.

3.8 Determine the power in the sidefrequencies of an amplitude-modulated waveform if the total power is 10 kW and the depth of modulation is 60%.

3.9 A 500 kHz carrier is amplitude modulated by signals in the frequency band of 250–3000 Hz. Determine the frequencies contained in the AM wave and state the required bandwidth.

3.10 Show that when a carrier at frequency f_c is amplitude modulated by a sine voltage at frequency f_m the bandwidth required is $B = 2f_m$ if $f_c > f_m$. Determine whether or not this relationship still applies if $f_m > f_c$.

3.11 A 50 kHz 12 V carrier is amplitude modulated by a 6 V sine wave at a frequency of 2500 Hz. Calculate the frequencies contained in the AM wave, the required bandwidth and the modulation factor.

3.12 A 500 kHz carrier is amplitude modulated by the band of frequencies 10–20 kHz. Determine the frequencies contained in (*a*) the lower sideband and (*b*) the upper sideband. What is the bandwidth necessary to transmit this signal?

3.13 A 28 V carrier wave is amplitude modulated to a depth of 50%. Draw the waveform and calculate the maximum and minimum voltages of the wave.

3.14 The envelope of a sinusoidally modulated AM wave has a maximum value of 12 V and a minimum value of 3 V. Calculate the voltages of (*a*) the carrier, (*b*) the modulating signal, and (*c*) the upper sidefrequency.

3.15 The voltage of the lower sidefrequency of a sinusoidally modulated AM wave is 6 V. If the carrier voltage is 24 V, calculate (*a*) the modulating signal voltage and (*b*) the modulation factor.

3.16 A carrier wave has an unmodulated power of 100 W. Calculate the power in the lower sidefrequency if the carrier is sinusoidally modulated to a depth of 40%.

3.17 A 1000 W carrier wave is amplitude modulated to a depth of 80%. What is the total power in the modulated wave?

3.18 Calculate the modulation factor of a radio system if the carrier voltage is 3 V and the signal voltage is 0.3 V.

3.19 Explain why the modulation index cannot be improved by amplifying the signal.

3.20 A 9 V carrier is amplitude modulated by a sinusoidal signal with the result that its amplitude varies up to 12 V. Calculate the modulation factor and the amplitude of each sidefrequency.

3.21 If the wave described in question 3.20 is applied across a 100 Ω resistor, calculate the power dissipated.

3.22 An 88 kHz carrier is amplitude modulated by a 0−4 kHz signal and the lower sideband is selected. This sideband is then used to amplitude modulate a 120 kHz carrier. Determine the frequencies contained in the lower sideband of the second modulation stage.

3.23 An AM waveform is applied across a 1000 Ω load. The powers produced by each of the sidefrequencies are 100 W and the carrier power is 2000 W. Calculate (*a*) the total power and (*b*) the modulation factor.

3.24 Calculate the peak voltage at the output of a circuit when the carrier voltage is 0.01 V and the signal voltage is 1 mV.

3.25 Draw the frequency-spectrum graphs of both of the following: (*a*) DSBAM and (*b*) SSBSC (upper sidefrequency). For both cases assume (i) carrier 12 V and 64 kHz, (ii) modulation signal 6 V and 0.3−3.4 kHz.

3.26 A 50 V carrier wave is amplitude modulated by a sinusoidal voltage. Calculate the voltage of (*a*) the modulating signal, (*b*) the upper sidefrequency and (*c*) the lower sidefrequency. The modulation depth is 25%.

3.27 An FM wave has a rated system deviation of 10 kHz. (*a*) What is the maximum frequency swing? (*b*) What will be the frequency swing when the modulation voltage is one-half its maximum permitted value?

3.28 An FM system has a rated system deviation of 6 kHz and a maximum modulation frequency of 3 kHz. Calculate the necessary bandwidth to transmit the signal.

3.29 A number of FM channels are to be transmitted over a wideband cable having a bandwidth of 500 kHz. Each channel has an audio bandwidth of 0−4 kHz and a rated system deviation of 7 kHz. Calculate how many channels can be transmitted.

3.30 A 30 MHz carrier is frequency modulated by a 15 kHz signal. The maximum frequency deviation of the carrier frequency is 70 kHz. Calculate (*a*) the modulation index and (*b*) the frequency swing of the modulated wave.

3.31 An FM system has a carrier frequency of 10 MHz. If the modulation frequency is 20 kHz and the available bandwidth is only 150 kHz, calculate (*a*) the deviation ratio and (*b*) the frequency swing.

3.32 An FM system has a rated system deviation of 6 kHz and a deviation ratio of 2. Calculate (*a*) the maximum modulation frequency and (*b*) the required bandwidth.

3.33 A 100 W carrier wave is frequency modulated and its modulation index is then 5. Determine the power in the FM wave.

3.34 An FM system has a maximum frequency deviation of 20 kHz. If the modulating signal voltage is 50% of the maximum permitted value, calculate the frequency deviation.

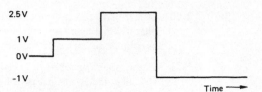

Fig. 3.18

3.35 Draw the FM wave that is produced by the modulating signal voltage waveform given in Fig. 3.18.

3.36 A 139 MHz carrier wave is frequency modulated by a 3 kHz wave. List the component frequencies in the modulated wave if up to third-order components are present.

3.37 A 142 MHz carrier is frequency modulated so that it has a modulation index of 6. If the modulation frequency is 10 kHz, calculate the frequency deviation of the carrier.

3.38 Explain the differences between the terms 'modulation index' and 'deviation ratio' and between 'frequency deviation' and 'rated system deviation'.

3.39 Compare FM and AM with respect to (a) the necessary bandwidth, (b) the power content and (c) the output signal-to-noise ratio.

4 Digital Modulation

The bandwidth of the commercial-quality speech circuit is 300–3400 Hz but at frequencies above about 3000 Hz group delay–frequency distortion increases to such an extent that data transmission becomes difficult. Because of this the highest frequency made available for data transmission is usually 3000 Hz. The public switched telephone network (PSTN) is unable to transmit signals at or near 0 Hz through line matching transformers, telephone exchange equipment, and multi-channel PCM systems. The direct transmission of data over a telephone circuit is only possible over *very* short lines, or over a digital circuit. For all other data links the data signal must be applied to a modem in which a carrier frequency can be modulated in amplitude, frequency or phase to produce an analogue voice-frequency signal that can be transmitted over the telephone network. As the use of the integrated service digital network (ISDN) becomes more common the need for modems will decrease. There are digital circuits now available to customers that are able to provide a high bit rate digital data circuit. These digital circuits are named *Kilostream* and *Megastream* by BT.

Modems

A *modem* is a piece of equipment that is fitted to each end of a data circuit to change the serial digital signals transmitted by the data terminal into voice-frequency signals suitable for transmission over the telephone line network. A modem also converts voice-frequency signals received from line into serial binary digital data signals which are then passed to the data equipment. A modem may be able to operate in one or more of the *simplex*, *half-duplex* and *full-duplex* modes.

Essentially, a modem has two parts: a transmitter and a receiver. The basic block diagram of the transmitting section of a modem is shown by Fig. 4.1. The encoder is needed for a modem working at bit rates of 2400 bits/s and higher when bits are grouped into twos (dibits), threes (tribits), or fours (quabits) (and sometimes fives (quinbits) or sixes (hexbits)). Modems operating at the lower bit rates of 300, 600 and 1200 bits/s do not use an encoder. The carrier frequency is applied to the modulator and is modulated by the data

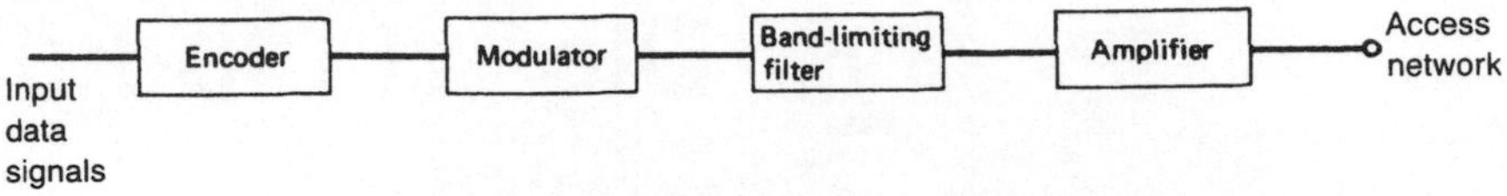

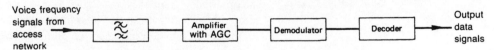

Input data signals — Encoder — Modulator — Band-limiting filter — Amplifier — Access network

Fig. 4.1 Transmitting section of a modem

signal to produce the corresponding voice-frequency signal. The bandwidth of the modulated data waveform is limited by the bandpass filter to restrict the transmitted signal to the bandwidth made available by the line. The band-limited signal is amplified before it is transmitted into the two-wire or four-wire line. The amplifier also ensures that the modem is impedance-matched to the line over the range of frequencies to be transmitted.

Figure 4.2 shows the basic block diagram of the receiver section of a modem. The incoming voice-frequency signals are filtered to remove unwanted noise and distortion components lying outside the wanted frequency band. The incoming signals are then amplified by the amplifier whose gain is controlled by an automatic gain control (AGC) circuit to ensure that the voltage input to the demodulator remains more or less constant at all times. The demodulator extracts the information content of the modulated waveform to produce a digital coded signal. If this signal has been coded into dibit, or tribit, etc., form it is then decoded to recover the original bit stream.

Voice frequency signals from access network — [filter] — Amplifier with AGC — Demodulator — Decoder — Output data signals

Fig. 4.2 Receiving section of a modem

Not all of the frequency spectrum of a commercial-quality speech circuit set up via the PSTN is available for the transmission of data since it was necessary to avoid those frequencies which were used for signalling over trunk circuits and for supervisory purposes. For this reason the available frequency spectrum for FSK signals (see p. 46) is restricted to 300–500 Hz and 900–2100 Hz, i.e. a bandwidth of 1200 Hz. Low-speed data links of up to 1200 bits/s can be operated so that each bit in the data waveform is represented by a single change in the modulated parameter. Higher-speed modems, operating at 2400 bits/s or more, must generally encode the data waveform before the modulation process is carried out. One method used is to pair bits together to form dibits, each dibit being used to produce a single change in the modulated carrier. The four possible dibits are 00, 01, 10 and 11. Since changes in the modulated carrier will now occur only *half* as often as when single bits are used, the baud speed on the transmission medium is reduced by half.

The bit grouping process can be carried out still further so that a single change in the modulated carrier occurs for every three bits (tribits), or every four bits (quabits), of information sent; the line baud speed will then be reduced threefold, or fourfold but at the expense of increased circuit complexity and hence increased cost. The tribits are 000, 001, 010, 011, etc. and the quabits are 0000, 0001, 0010, etc. The data stream may even be encoded into fives to form the quinbits 00000, 00001, 00010, etc. Even six-bit encoding (hexbits) is used for 14 400 trellis-encoded systems.

Example 4.1

Data signals are transmitted over a circuit at a line baud speed of 2400 bauds. Calculate the bit rate if (*a*) each bit is transmitted separately, or the data is encoded into (*b*) dibits (*c*) tribits or (*d*) quabits.

Solution
(*a*) Bit rate = baud speed = 2400 bits/s   (*Ans.*)
(*b*) Bit rate = 2400 × 2 = 4800 bits/s   (*Ans.*)
(*c*) Bit rate = 2400 × 3 = 7200 bits/s   (*Ans.*)
(*d*) Bit rate = 2400 × 4 = 9600 bits/s   (*Ans.*)

Most modems conform to one of the ITU−T V recommendations listed in Table 4.1. It can be seen that *frequency shift modulation* (FSK) is only employed at low bit rates of up to 1200 bits/s. Higher bit rates employ either *differential phase shift modulation* (DPSK) or *quadrature amplitude modulation* (QAM).

V22 bis is currently regarded as the basic entry system; it is commonly used in conjunction with MNP (Microcom Network Protocol) 4 and 5 which are, respectively, error detection and data compression schemes. At the higher bit rates two ITU−T

Table 4.1

ITU−T recommendation	Modulation method	Bit rate (bits/s)
V21	FSK	300
V22	DPSK	1200/600/300
V22 bis	QAM	2400/1200
V23	FSK	1200/600
V26	DPSK	2400
V26 bis	DPSK	2400
V26 ter	QAM	2400/1200
V27	DPSK	4800
V27 bis	DPSK	4800/2400
V27 ter	DPSK	4800/2400
V29	QAM	9600/7200/4800
V32	QAM	9600/4800/2400
V32 bis	QAM	14400
V32 terbo	QAM	19200 (over all-digital line)
V33	QAM	14400/12000
V34	QAM	28800/26400/21600 (etc. downwards in 2400 bits/s increments to 2400 bits/s)

recommendations, namely V42 and V42 bis, are often used; V42 is an error correction system which makes a data link more reliable, and V42 bis is a method of data compression. A modem that uses V42/V42 bis can (in theory) quadruple the speed of V32 bis to 57 600 bits/s. PCs, however, have a maximum output speed of only 38 400 bits/s. The use of V32 with MNP5 data compression gives a bit rate of 19 200 bits/s.

Frequency Shift Modulation

When a sinusoidal carrier wave is frequency modulated, its frequency is made to vary in accordance with the characteristics of the modulating signal. The amount by which the carrier frequency is deviated from its nominal value is *proportional* to the amplitude of the modulating signal, and the number of times per second the carrier frequency is deviated is *equal* to the modulating frequency.

When the modulating signal is a data waveform, the amplitude of the modulating signal is fixed at a particular value, say ± 6 V, and its maximum fundamental frequency is one-half of the bit rate. This means that the carrier frequency is *always* deviated to either one of two different frequencies; the *nominal* carrier frequency is the arithmetic mean of these two frequencies but it is never actually present since the modulating signal is never at 0 V. The higher of the two frequencies is used to represent binary 0 while the lower frequency represents binary 1. The system is usually known as FSK.

The higher the bit rate of the data signal the greater must be the separation between the two frequencies representing binary 1 and binary 0. Otherwise, the detector in the modem receiver will be unable to determine reliably which of the two frequencies is present at any particular instant in time. Because of this and because of the limited bandwidth made available by the PSTN the maximum bit rate provided by an FSK system operating over telephone circuits is 1200 bits/s.

V21 FSK

For bit rates of 300 bits/s it is possible to accommodate two channels on a two-wire presented speech circuit and thus provide duplex transmission. The frequencies used to represent the bits 1 and 0 for the standard bit rates are given by Table 4.2. In each case the nominal carrier frequency is the arithmetic mean of the two frequencies.

An example of an FSK waveform is given by Fig. 4.3. When the

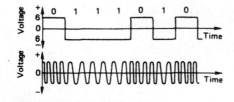

Fig. 4.3 Frequency-shift waveform

Table 4.2 V21 FSK

Bit rate (bits/s)	Frequencies (Hz)		
	Binary 0	Binary 1	
up to 300	1180	980	(different directions
	1850	1650	of transmission)

data waveform is at $+6$ V (binary 0), the transmitted signal is at the higher of the two frequencies and, similarly, the lower frequency is transmitted to represent binary 1.

V21 FSK is now rarely used for data communication because of its low speed but it is still employed to transmit the procedural messages for group 3 FAX machines.

V23 FSK

Table 4.3 V23 FSK

Bit rate	Frequencies (Hz)	
	Binary 0	Binary 1
600	1700	1300
1200	2100	1300

V23 FSK provides half-duplex operation over a connection set up via the PSTN or full-duplex operation over a dedicated leased circuit. The bit rate is either 600 bits/s or 1200 bits/s. The frequencies employed to indicate binary 1 and binary 0 are given by Table 4.3.

Operation at 300 bits/s is full-duplex over the PSTN but only half-duplex at 600 bits/s and 1200 bits/s.

Modulation Index and Deviation Ratio

The modulation index m_f and the deviation ratio D of a frequency-modulated wave were defined in Chapter 3 (Equations 3.8 and 3.9).

In the case of frequency shift modulation, the frequency deviation is fixed since the amplitude of the modulating signal voltage — the data waveform — is constant, often ± 6 V. The maximum modulating frequency exists when the data waveform consists of alternate 1s and 0s and it is then equal to one-half of the bit rate. Thus the deviation ratio of a frequency shift system is given by

$$D = \text{(tone separation)/(bit rate)} \qquad (4.1)$$

Example 4.2

Calculate the deviation ratio for a 1200 bits/s frequency-shift data system.

Solution
For a 1200 bits/s data system the two frequencies used are 1300 Hz and 2100 Hz.

Thus, the deviation ratio $= (2100 - 1300)/1200 = 0.67$ *(Ans.)*

In this calculation of the deviation ratio the harmonics of the data waveform have not been considered. If this calculation is repeated for the other standard bit rate it will be found that the same answer is obtained.

The Frequency Spectrum of a Frequency-shift Waveform

When a sinusoidal carrier wave of frequency f_c is frequency modulated by a sinusoidal signal of frequency f_m, the modulated wave will contain components at a number of different frequencies.

The bandwidth needed for the transmission of a frequency-modulated wave is given by

$$\text{Bandwidth} = 2(f_d + f_m) \tag{4.2}$$

where f_d is the maximum frequency deviation and f_m is the maximum modulating frequency.

For a frequency-shift waveform the bandwidth expression can be rewritten as

$$\text{Minimum bandwidth} = 2(f_d + \text{bit rate}/2) \tag{4.3}$$

$$= \text{tone separation} + \text{bit rate} \tag{4.4}$$

Suppose a 1200 bits/s frequency-shift waveform is to be transmitted. From equation (4.4) the bandwidth needed is

$$\text{Bandwidth} = (2100 - 1300) + 1200$$

$$= 2000 \, \text{Hz}$$

This means that the bandwidth needed to transmit a 1200 bits/s frequency-shift system is greater than the bandwidth that is available (900–2100 Hz or 1200 Hz) when a data system is set up over the PSTN. Fortunately, it is not necessary that the waveform of the signal arriving at the far end of a link is undistorted. It is sufficient that the receiving equipment is able to determine at any instant whether a binary 1 or binary 0 bit is being received. This means that it is *not* necessary for all the significant sidefrequencies to be transmitted. Since the deviation ratio of a frequency-shift waveform is approximately unity, most of the energy content of the waveform is concentrated in the carrier and the first-order sidefrequency components. Hence, *only* these components need to be received by the FSK receiver.

The advantage of transmitting only the first-order sidefrequencies is the considerable reduction in the occupied bandwidth that is obtained. The minimum bandwidth necessary is then only *twice* the maximum modulating frequency or

$$\text{minimum bandwidth} = \text{bit rate} \tag{4.5}$$

In the previous example of a 1200 bits/s frequency-shift system the necessary bandwidth is now only 1200 Hz as opposed to the 2000 Hz previously calculated. This narrower bandwidth can be accommodated in the frequency spectrum made available by the PSTN (900–2100 Hz).

The basic block diagram of an FSK modulator is shown by Fig. 4.4. The data waveform to be transmitted is band-limited by the input bandpass filter which has a cut-off frequency of 1300 Hz. The waveform applied to the voltage-controlled astable multivibrator consists of the fundamental frequency only of a 1200 bits/s signal or the fundamental plus third harmonic of a 600 bits/s signal. The voltage-controlled multivibrator is switched between two states: one in which

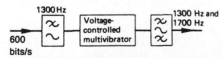

Fig. 4.4 FSK modulator

it oscillates at 1300 Hz and one in which it oscillates at either 1700 Hz or 2100 Hz (depending on the bit rate). The square output waveform of the multivibrator is applied to a bandpass filter whose bandwidth is sufficiently narrow to ensure that only the fundamental frequency of either 1300 Hz or 1700 Hz (or 2100 Hz) is passed. The output of the filter is thus the wanted frequency shift waveform.

Phase Shift Modulation

Frequency shift modulation is only available for bit rates up to 1200 bits/s since the next higher standard bit rate of 2400 bits/s would need a bandwidth of 2400 Hz which cannot be provided by the PSTN. For 2400 bits/s systems therefore, *phase shift modulation* is used.

When a sinusoidal carrier is phase modulated, its instantaneous phase is made to vary in accordance with the characteristics of the modulating signal. The magnitude of the phase deviation is *proportional* to the modulating signal voltage and the number of times per second the phase is deviated is *equal* to the modulating frequency. In a data system the modulating signal voltage is ± 6 V and so the phase deviation obtained is fixed.

Modulating the phase of the carrier will at the same time vary the instantaneous carrier frequency since angular velocity ($\omega = 2\pi \times$ carrier frequency) is the rate of change of phase.

The modulation index of a phase-modulated waveform is, as with frequency modulation, the maximum phase deviation of the carrier frequency produced by the modulating signal. However, the value of the modulation index depends only upon the modulating signal voltage and is quite independent of the modulating frequency.

The modulation process generates a number of sidefrequencies spaced symmetrically either side of the carrier frequency, and the frequency spectrum is exactly the same as that of a frequency-modulated wave having the same numerical value of modulation index.

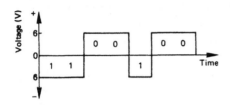

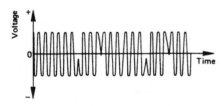

Fig. 4.5 Phase-modulated waveform

An example of the use of phase modulation is given by Fig. 4.5. The phase of the carrier is shifted by 180° each time the leading edge of a 0 bit occurs. The phase of the carrier is not altered by a 1 bit. In practice, phase modulation is rarely employed, because its detection is difficult, and instead *differential phase modulation* (DPSK) is employed. This version of phase modulation uses *changes* in phase, rather than phase itself, to indicate the dibits or tribits into which a data stream has been encoded.

Differential Phase Shift Modulation

V22, V26 and V26 bis DPSK

A DPSK system that operates at 1200 bits/s or at 2400 bits/s encodes the data bit stream into dibits that represent 00, 01, 11 and 10. Each dibit is represented by a different *phase change* (not an absolute value of phase). The phase changes used for these three systems are given by Table 4.4.

Table 4.4 V22, V26, V26 bis DPSK

		Dibit			
		00	*01*	*11*	*10*
Phase change	V22	+90°	+0°	+270°	+180°
	V26	+0°	+90°	+180°	+270°
	V26 bis	+45°	+135°	+225°	+315°

The use of dibits reduces the baud speed on the line and hence the bandwidth that must be made available since the line signal changes state only half as often as it would if all bits were directly transmitted to line.

When dibits are used, the equipment at the receiving end of the system must be synchronized with the transmitting equipment for decoding to take place correctly. The necessary synchronization is developed from the changes in phase of the received signal.

Suppose, for example, that the data signal 10100111 is to be transmitted (least significant bit (LSB) first). The bits are grouped together in the decoder section of the modem to form the dibits 11, 01, 10, 10. These dibits are then signalled to line by changing the phase of the carrier by, in turn, +225°, +135°, +315°, +315°. The carrier frequency used is 1800 Hz since at this frequency the group delay—frequency distortion of a line is small and the frequency is approximately at the middle of the available frequency spectrum.

A V22 or V26 DPSK signal is often represented by a *constellation diagram* that shows the amplitude and phase relationships of the dibits transmitted to line. The constellations of the three systems are shown by Fig. 4.6(*a*), (*b*) and (*c*). Each point in a constellation represents a dibit of information. Figure 4.6(*c*) shows the constellation of a V26 bis DPSK signal. The dibit 00 is represented by a phase change of 45° so it is indicated by a point positioned at the mid-angle of the first quadrant, 0° to 90°, of the constellation diagram. Dibit 01 is represented by a 135° phase change so it has a point in the second, 90° to 180°, quadrant. Similarly, dibits 11 and 10 are represented by points in the third and the fourth quadrants respectively.

A 1800 Hz carrier is used to position the DPSK signal in the part of the available bandwidth where group delay—frequency distortion is least. The modulation process produces frequency components located both above and below the carrier frequency, each component below 1800 Hz having a corresponding component above 1800 Hz. Each pair of components is separated by a frequency gap numerically equal to the line baud speed. Because the bit rate is 2400 bits/s and there are two bits per dibit the line baud speed is 1200 bauds.

The minimum bandwidth needed to transmit a DPSK signal is given by

$$\text{Minimum bandwidth} = (\text{bit rate})/\log_2 n$$
$$= (\text{bit rate})/(3.32 \log_{10} n) \qquad (4.6)$$

where n is the number of phase changes employed.

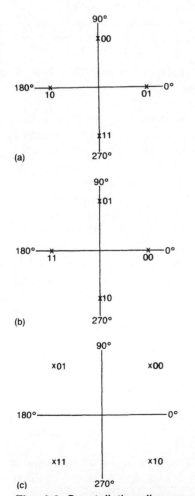

Fig. 4.6 Constellation diagrams: (*a*) V22 DPSK, (*b*) V26 DPSK and (*c*) V26 bis DPSK

Example 4.3

Calculate the minimum bandwidth needed for a V26 DPSK signal.

Solution
$n = 4$
Minimum bandwidth $= 2400/(3.32 \log_{10} 4) = 1200$ Hz   (*Ans.*)

V22 can give either half-duplex or full-duplex operation over either the PSTN or a leased circuit. The carrier frequencies used are 1200 Hz in one direction of transmission and 2400 Hz in the other direction.

V27 DPSK

For operation at 4800 bits/s the data stream is encoded into the tribits 000, 001, 010, etc. Eight different tribits are possible and so eight phase changes are necessary to represent them. The V27 phase changes are given by Table 4.5. The line baud speed is 4800/3 = 1600 bauds and the minimum bandwidth necessary is, from equation (4.6),

$$\text{minimum bandwidth} = 4800/(3.32 \log_{10} 8) = 1600 \text{ Hz}$$

Table 4.5 ITU–T V27 DPSK

Tribit	000	001	010	011	100	101	110	111
Phase change	+45°	+0°	+90°	+135°	+270°	+315°	+225°	+180°

[Note that once again the minimum bandwidth is equal to the line baud speed].

The constellation diagram of a 4800 bits/s V27 DPSK signal is shown by Fig. 4.7(*a*). V27 bis and V27 ter operate at 4800 bits/s with the same phase change values to represent different tribits; their constellation diagram is shown in Fig. 4.7(*b*). V27 gives full-duplex

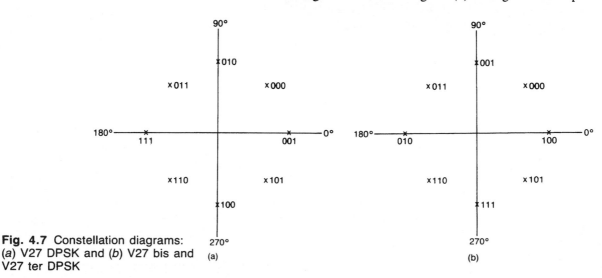

Fig. 4.7 Constellation diagrams: (*a*) V27 DPSK and (*b*) V27 bis and V27 ter DPSK

operation over a leased circuit; V27 bis gives half-duplex working over the PSTN and full-duplex over a dedicated circuit, while V27 ter can give just half-duplex service over the PSTN.

Quadrature Amplitude Modulation

Although, in theory, it is possible to extend the concept of DPSK to quabits it would require $2^4 = 16$ phase changes and this would be difficult to implement. Instead, higher bit rates than 4800 bits/s are obtained with the use of *quadrature amplitude modulation* (QAM). QAM is a combination of amplitude modulation and phase modulation that is employed at bit rates of 2400 bits/s and higher. The data stream to be transmitted is divided up into quabits and two carriers are employed that are at the same frequency but are 90° out of phase with one another. When one of the carriers is at its maximum amplitude the other carrier is at zero volts.

Thus the equation for the instantaneous voltage of the combined carrier signal is

$$v = V_1 \sin\omega\, t + V_2 \cos\omega\, t \tag{4.7}$$

V22 bis and V22 ter QAM

In both these systems the data stream is encoded into quabits; two bits are used to change the phase of the carrier and two bits are used to change its amplitude. This gives a total of $2^2 = 4$ phase changes and 4 different amplitudes. The line baud speed is 2400/4 = 600 bauds. This allows full-duplex operation over a dial-up PSTN circuit or a dedicated circuit; in one direction of transmission the carrier frequency employed is 1200 Hz and in the other direction the carrier frequency is 2400 Hz.

The most significant dibit selects one of the four quadrants of the constellation diagram; dibit 00 selects the first quadrant, dibit 01 selects quadrant 2 and dibits 11 and 10 select the third and fourth quadrants, respectively. This is shown by Fig. 4.8(*a*). The least significant dibit selects one of four points in the selected quadrant as shown by Fig. 4.8(*b*). The points in the first quadrant have the relative amplitudes given in Table 4.6.

Points in the other quadrants have similar amplitudes but rotated through 90° for each quadrant.

Table 4.6 V22 bis, V22 ter QAM

Dibit	Horizontal direction	Vertical direction
00	1/3	1/3
01	1	1/3
11	1	1
10	1/3	1

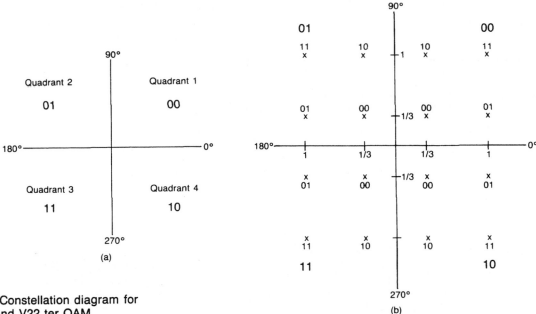

Fig. 4.8 Constellation diagram for V22 bis and V22 ter QAM

Example 4.4

Determine the equation for the combined carrier voltage when the quabit (*a*) 0010, (*b*) 1010 and (*c*) 1111 is transmitted.

Solution
(a) $v = 1/3 \cos \omega t + 1 \sin \omega t$ (*Ans.*)
(b) $v = 1/3 \cos \omega t - 1 \sin \omega t$ (*Ans.*)
(c) $v = -1 \cos \omega t - 1 \sin \omega t$ (*Ans.*)

V22 bis and ter are often used in conjunction with the Microcon Network Protocol (MNP) 4 and 5; MNP 4 gives error protection and MNP 5 is a data compression standard.

V29 QAM

The V29 modulation system uses quabits in which the last three bits are used to give eight different phases, as in the V27 DPSK system (see Table 4.5), and the first bit gives different amplitudes. The phases and amplitudes used for V29 are listed in Table 4.7.

The constellation diagram for a V29 signal is shown in Fig. 4.9(*a*). There are 16 points shown on the diagram. Table 4.8 shows how each phase is selected by a particular tribit.

In the 7200 bits/s fall-back mode the first bit is 0 at all times and the other three bits keep the same function as at 9600 bits/s. The 8-point constellation diagram is shown by Fig. 4.9(*b*).

Table 4.7 V29 QAM

Phase	First bit	Relative amplitude
0°, 90°, 180°, 270°	0	3
	1	5
45°, 135°, 225°, 315°	0	$\sqrt{2}$
	1	$3\sqrt{2}$

Table 4.8 V29 QAM

Phase change	Bits		
	2	3	4
0°	0	0	1
45°	0	0	0
90°	0	1	0
135°	0	1	1
180°	1	1	1
225°	1	1	0
270°	1	0	0
315°	1	0	1

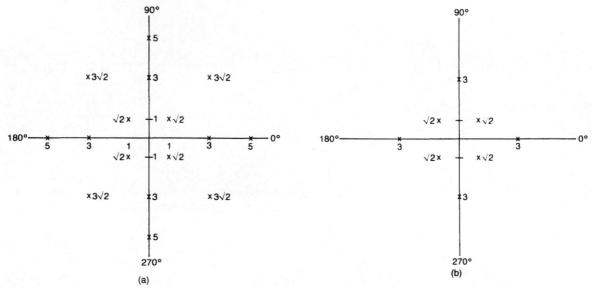

(a)   (b)

Fig. 4.9 Constellation diagrams for V29 QAM: (a) 9600 bits/s, (b) 7200 bits/s

A V29 modem can operate at 9600 bits/s or 7200 bits/s or, rarely, 4800 bits/s over either the PSTN or a dedicated circuit. If the line is two-wire presented only half-duplex operation is possible, but a four-wire presented line allows full-duplex working.

V32 QAM

The V32 QAM method of digital modulation uses quabits to provide 9600 bits/s over a 2400 baud circuit via the PSTN. The two most significant bits are used to select one of the four quadrants and the two least significant bits select either of two possible amplitudes. The constellation diagram is shown in Fig. 4.10. It can be seen that the most significant pair of bits 00 selects the third quadrant, dibit 01 selects quadrant 4, quadrant 1 is selected by the pair of bits 11 and quadrant 2 is selected by dibit 10. The least significant dibits select one of the four possible amplitude choices from 1/3 and 1, i.e. 1/3 and 1/3, 1 and 1/3, 1/3 and 1, and 1 and 1.

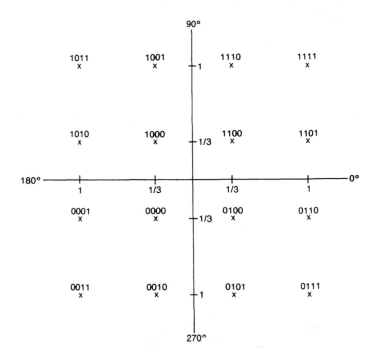

Fig. 4.10 Constellation diagram for
V32 QAM

V32 Trellis-encoded QAM

At bit rates of 9600 bits/s, or even higher rates such as 14 400 bits/s
and 19 800 bits/s, line distortion ensures that it can be difficult to obtain
reliable reception of data. To obtain reliable reception *trellis encoding*
is frequently employed. This consists of the use of a fifth encoding
bit that is added to each quabit, giving a total of $2^5 = 32$ different
bit combinations or quinbits. The code bit is chosen so that only certain
specified bit sequences are allowable. When the transmitted data
arrives at the receiver the receiver matches the received signal with
the 32-point trellis-coded pattern. Any difference that is detected is
an indication that an error has occurred in transmission. The
constellation diagram for a V29 trellis-encoded signal is shown in Fig.
4.11.

V33 QAM

V33 QAM operates full-duplex at 14 400 bits/s over a four-wire
presented circuit with a fall-back speed of 12 000 bits/s. It uses septbits
to obtain a line baud speed of 2400 bauds with trellis encoding. The
constellation diagram is very complex (128 points), as can be seen
from Fig. 4.12.

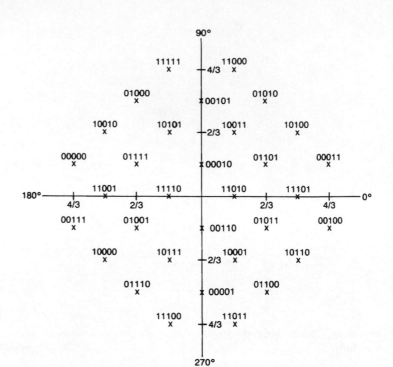

Fig. 4.11 Constellation diagram for
V32 trellis-encoded QAM

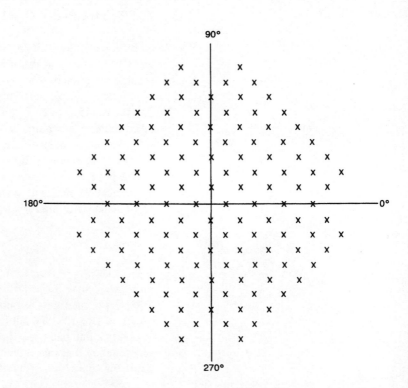

Fig. 4.12 Constellation diagram for
V33 QAM

The Effects of Lines on Modulated Data Signals

When modulated data signals are transmitted over a line, attenuation and group delay—frequency distortion may distort the received signal to such an extent that the receive modem may not be able to interpret the incoming signals correctly. The bit error rate will then rise. The effects of the line characteristics will be explained by considering frequency shift modulation.

When frequency shift modulation is used, the higher frequency transmitted will be attenuated to a greater extent than the lower frequency and hence the 0 bits may become of so low a level that it becomes difficult for the receiver to determine when a 0 bit has been received. This effect is much more prevalent for 1200 bits/s systems than for 600 bits/s systems because the former system uses the higher frequency of 2100 Hz to represent a 0 bit. Thus the effect of line attenuation is to reduce the maximum data transmission rate. The effect of group delay—frequency distortion is to delay the 2100 (or 1700) Hz pulses to a greater extent than 1300 Hz pulses, so that the higher frequency pulses arrive at the end of the link at incorrect instants in time (relative to the lower frequency bits). The effect of this is to make the pulses overlap so that binary 1 bits appear longer and binary 0 bits appear shorter than they should. In extreme cases the receive modem may not be able to detect a 0 bit received in between two 1 bits.

Similar but smaller effects occur when DPSK or QAM are used because the changes in phase are signalled to line by various pairs of frequencies. Modems operating at higher bit rates incorporate adaptive equalizers to optimize the overall line—modem amplitude and phase characteristics. QAM is used at the higher bit rates because, along with much more complex modem circuity, it is less affected by the characteristics of the transmission path.

Exercises

4.1 (a) Outline a method by which data may be transmitted over an audio-frequency circuit at 2400 bauds. (b) How can errors in the received data be detected?

4.2 Explain, with the aid of waveform sketches, what is meant by frequency shift modulation. A 1500 Hz carrier wave is frequency-shift modulated by a 600 bits/s data waveform.
(a) What is the maximum fundamental frequency of the data waveform?
(b) What bandwidth is required for the transmission of the frequency-shift waveform if (i) the fundamental component only is transmitted, (ii) the third harmonic is also transmitted?
(c) Why is it possible to transmit only the first-order sidefrequencies and what is the advantage gained by so doing?

4.3 What is meant by the term *dibit* used in data communications? How can dibits be represented by (a) changes in the phase angle of a carrier wave, (b) changes in the amplitude of a carrier wave? State the advantage which is gained by the use of multi-phase or multi-level transmission.

4.4 Draw the waveform of a carrier wave that has been frequency shift modulated by the data signal 1011001 at a bit rate of 1200 bits/s. Label the diagram with the frequencies that are transmitted and state the nominal carrier frequency. Which data waveform will have the highest fundamental frequency and what is then the occupied bandwidth?

4.5 Explain how the performance of a frequency-shift system can be adversely affected by line attenuation and group delay—frequency distortion. Why is a 1200 bits/s system more likely to be affected than a 600 bits/s system?

4.6 Draw the block diagram of a modem, showing both transmitting and receiving sections, and briefly explain its operation. Discuss the factors that influence the choice of carrier frequency.

4.7 What is the bandwidth of a commercial-quality speech circuit? Why cannot all of this bandwidth be employed for data transmission? State what bandwidth is available (*a*) for an FSK system, (*b*) for a DPSK system.

4.8 How does frequency shift modulation differ from frequency modulation? Illustrate your answer with waveform sketches. Calculate the deviation ratio for a 600 bits/s frequency-shift data system. What bandwidth is needed for such a system?

4.9 The bit stream 1010010111100011 is to be encoded using (*a*) dibits, (*b*) tribits, (*c*) quabits and (*d*) quinbits. Divide the bit stream into groups and, selecting and stating suitable amplitudes and phases to represent each group, draw the constellation diagrams.

4.10 Figure 4.13 shows a constellation diagram. Identify it. State the bit rates for which this sytem is employed. Also state the line baud speed.

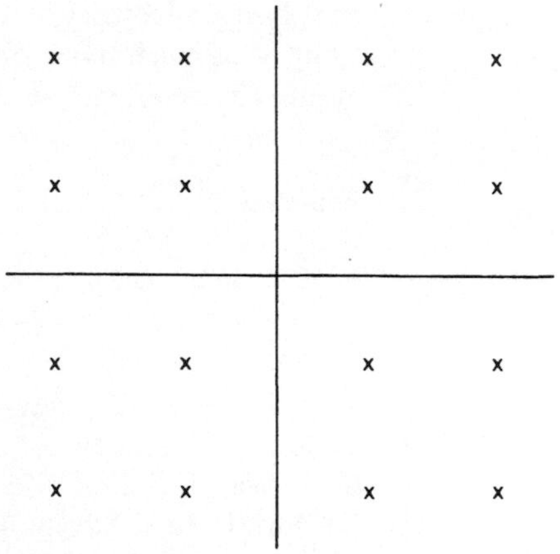

Fig. 4.13

5 Carrier Frequencies and Bandwidths

A number of different methods of modulations are employed in telecommunication engineering to enable the best use to be made of the frequency spectrum made available by a given transmission medium. For each channel in a communication system a *carrier frequency* must be chosen to position the channel in the required part of the frequency spectrum. To obtain the maximum utilization of the transmission medium and to minimize costs, the channel bandwidth must be as narrow as possible; it must, however, be wide enough to pass all the significant frequency components of the signal. The signals to be transmitted over the various types of communication system fall into one of four classes: (*a*) telephony, (*b*) music, (*c*) television and (*d*) data.

Line Communication Systems

Telephony

Telephone cables are capable of transmitting a band of frequencies well in excess of the normal speech frequency range and can therefore be used to carry single-sideband amplitude-modulated frequency-division multiplex telephony systems. A number of channels are made available by such a system, each channel being allocated a different carrier frequency. The range of carrier frequencies depends upon the number of channels provided by the system and the part of the frequency spectrum which the system is to occupy.

The number of channels that can be carried by a single cable pair is primarily determined by the attenuation (loss) of the cable at the highest frequency to be transmitted. The greater the cable attenuation the closer together must the line amplifiers be spaced in order to prevent the signal level falling below a predetermined value. FDM multi-channel telephony systems are no longer employed in the trunk network of the UK.

The channel carrier frequencies for a 12-channel FDM system are specified by the ITU–T and are listed in Table 5.1 The table also gives details of the passband of each channel filter; it should be noted

Table 5.1 ITU−T FDM carrier frequencies

Channel no.	Carrier frequency (kHz)	Channel filter passband (kHz)
1	108	104.6−107.7
2	104	100.6−103.7
3	100	96.6− 99.7
4	96	92.6− 95.7
5	92	88.6− 91.7
6	88	84.6− 87.7
7	84	80.6− 83.7
8	80	76.6− 79.7
9	76	72.6− 75.7
10	72	68.6− 71.7
11	68	64.6− 67.7
12	64	60.6− 63.7

the bandwidths correspond to an audio bandwidth of 300−3400 Hz.

The transmitted bandwidth is therefore 60.6−107.7 kHz, or approximately 60−108 kHz.

PCM Systems

The minimum bandwidth required to transmit a PCM signal is equal to one-half of the bit rate used. For the basic 30-channel ITU−T telephony system the line bit rate is 2.048 Mbits/s so that the minimum bandwidth that must be provided is 1.024 MHz.

Telegraphy

FAX signals are transmitted using V29 QAM modulation at 9600/7200 bits/s or V27 ter at 4800/2400 bits/s. V29 modulation uses a line baud speed of 2400 and a carrier frequency of 1700 Hz. The bandwidth required is from 500 Hz to 2900 Hz. V27 ter uses DPSK modulation; at 4800 bits/s it has 1600 baud line speed, a carrier frequency of 1800 Hz, and a bandwidth of from 1000 Hz to 2600 Hz; at 2400 bits/s the bandwidth is reduced to 1200 to 2400 Hz.

Group 3 terminals currently dominate the FAX market and these transmit the information at either 2400 bits/s or 4800 bits/s using FSK. The standard frequencies employed are either 1300 ± 500 Hz or 1900 ± 550 Hz.

Radiocommunication Systems

The radiocommunication systems in use today may be divided into one of three main classes: (*a*) *broadcasting* — both radio and

Table 5.2

Frequency band	Classification	Abbreviation
10–30 kHz	very low frequencies	VLF
30–300 kHz	low frequencies	LF
300–3000 kHz	medium frequencies	MF
3–30 MHz	high frequencies	HF
30–300 MHz	very high frequencies	VHF
300–3000 MHz	ultra high frequencies	UHF
3–30 GHz	super high frequencies	SHF
30–300 GHz	extra high frequencies	EHF

television (*b*) *radio links* for providing point-to-point telephonic communication and (*c*) mobile radio.

Table 5.2 gives the classification of the various frequency bands used in radiocommunication.

Sound Broadcasting

A sound broadcasting system is one in which programmes of mixed entertainment, news and educational content are made available to a large number of people with radio receivers. In the LF band a single transmitter can serve the whole country, but at higher frequencies in the MF band two, or more, transmitters are required for countrywide service. At even higher frequencies in the VHF band the range of a particular station is very much smaller and it becomes necessary to have a number of transmitting stations located in different parts of the country.

The BBC broadcast their Radio 1, 2, 3, 4 and 5 programmes at a number of different frequencies in the MF and VHF bands and at one frequency in the LF band. In the MF band, amplitude-modulated DSB transmissions are used with carrier frequencies in the range 647–1546 kHz and having a bandwidth of approximately 9000 Hz. In Europe, by international agreement, carrier frequencies in the MF band are spaced at 9000 Hz intervals and, consequently, the highest audio frequency transmitted is in the region of 4500 Hz, i.e. a bandwidth of 9000 Hz.

The use of a 4.5 kHz audio bandwidth for MF broadcast transmissions results in poor reproduction of music because many of the higher harmonics produced by the musical instruments are suppressed. For reasonably high-quality reception of music, an audio bandwidth of at least 15 kHz is required. This occupies an r.f. bandwidth of 30 kHz, which cannot be accommodated in the congested MF band. High-quality broadcast transmissions are therefore provided in the VHF band using carrier frequencies in the range of 88.1–96.8 MHz. VHF signals have a limited range of about a hundred kilometres and so a number of stations are required within a fairly small area. The carrier frequencies allocated to the stations in a given area are about 200 kHz apart to minimize inter-station interference. The wide carrier frequency spacing allows frequency modulation to be used; this gives a better signal-to-noise ratio than amplitude modulation but it requires a much wider bandwidth (in the BBC VHF broadcast system an r.f. bandwidth of 180 kHz is necessary to provide the 15 kHz audio bandwidth).

Sound broadcast transmissions are also radiated in certain parts of the high-frequency band:

5.950—6.20 MHz, 7.1—7.3 MHz, 9.5—9.775 MHz,
11.7—11.975 MHz, 15.1—15.45 MHz, 17.7—17.9 MHz,
21.45—21.75 MHz and 25.6—26.1 MHz,

with a bandwidth of 10 kHz. Transmissions in these frequency bands are employed in Europe for international broadcast programmes as, for example, BBC programmes to Europe and to North Africa.

Television Broadcasting

Four television services are available to the British viewing public; two are provided by the BBC, the third, Channel 3, is provided by various commercial television companies, and the fourth is Channel 4. Channels 3 and 4 are transmitted by the Independent Broadcasting Authority (IBA). Colour and monochrome television signals are transmitted in the UHF band by both the BBC and the IBA. These programmes use vestigial sideband amplitude modulation for the picture signal, and frequency modulation for the sound signal. The vision carrier frequencies used are in the bands 471.25—575.25 and 615—847.25 MHz, providing a video bandwidth of 5.5 MHz. The sound carrier frequencies are spaced 6.0 MHz above the associated vision carrier and provide an audio bandwidth of 20 kHz.

Point-to-Point HF Radio Systems

HF radio links are used to route telephone calls to overseas destinations where circuits via submarine cable and/or communication satellites are not available or have inadequate capacity. Different carrier frequencies are employed for each direction of transmission; and the receiving stations are geographically situated well apart to minimize interference between signals passing in different directions. Radio links may be used to carry ordinary telephone conversations or to carry facsimile telegraphy or press broadcasts. The majority of radio links have carrier frequencies in the HF band and generally use a form of SSB amplitude modulation that is known as *independent sideband*. Radio-telegraphy services, such as the Press Broadcast Service, which is provided for the use of news agencies, are also generally provided in the HF band.

Several frequency bands have been allocated to HF radio links; some of these, for example, are

3.5–3.9 MHz, 5.73–5.95 MHz, 9.04–9.5 MHz,
13.36–14 MHz, 21.75–21.87 MHz, 26.1–27.5 MHz.

The bandwidth per channel is 250 Hz–3 kHz.

Mobile Systems

Maritime

Telephonic and telegraphic services to ships at sea are provided at a number of specified frequencies in the MF, HF, and VHF bands. Short-range telephony services, up to about 80 km in distance, are operated at frequencies in the VHF band, 156–163 MHz. Telephony services to ships at distances in excess of 80 km and less than about 1000 km are provided on specified carrier frequencies in the band 1.6–3.8 MHz. Ship-to-shore telegraphy services are operated, using the Morse code, in the band 405–525 kHz, except for some ships in northern seas which may use specified channels in the 1.6–3.8 MHz band. Longer distance services to ships at sea are operated at the following frequencies in the HF band; 4, 6, 8, 13, 17 and 22 MHz. 500 kHz (telegraphy) and 2182 kHz (telephony) are used as international distress and calling signals.

Radio communication services to ships at sea are increasingly provided by communication satellite by a service known as INMARSAT. It uses the frequency band 1.5 GHz ship-to-satellite and 1.6 GHz satellite-to-ship. The frequencies employed for satellite-to-shore links are in the 6/4 GHz band. Frequency modulation is used.

Land

Land mobile services are divided into three categories: (*a*) *emergency* (ambulance, fire and police), (*b*) *public utilities* (electricity, gas, telecoms and water) and (*c*) *private* (mini-cabs, taxis, delivery vans, etc.). Collectively, these services are known as the *private land-mobile* (PMR) services since they do not have access to the public switched telephone network (PSTN).

A considerable number of different frequency bands, of various widths, have been allocated to land, sea and air mobile services in the VHF and UHF bands. The complete list of frequencies is too long to include in this book and so Table 5.3 gives some examples.

Public land-mobile services are linked to the PSTN; a user can make a call to a PSTN customer or, alternatively, a mobile telephone may be called from a local telephone exchange line. Car telephones use a system known as *cellular radio* in which the country has been divided

Table 5.3

Frequency band (MHz)	Used by	Channel bandwidth (kHz)
71.5– 78.0	PMR	12.5
80.0– 85.0	emergency	12.5
85.0– 88.0	PMR	12.5
97.0–102.0	emergency	12.5
105.0–108.0	PMR	12.5
108.0–136.0	aero	25
138.0–141.0	PMR	12.5
146.0–148.0	emergency	12.5
156.0–163.0	maritime	25
165.0–173.0	PMR	12.5
425.0–449.5	PMR	12.5
451.0–452.0	emergency	25
453.0–462.5	PMR	25
465.0–466.0	emergency	25

up into a large number of geographically small areas known as *cells*. The mobile telephones transmit on a channel in the frequency band 890 to 935 MHz and receive on another channel in the frequency band 935 to 980 MHz. Another system uses the frequency bands 872 to 904 MHz and 917 to 950 MHz. Frequency modulation is employed with a peak frequency deviation of 9.5 kHz and a channel bandwidth of 25 kHz. A European system, known as GSM, will allow customers to use the same terminals and procedures all over Europe. GSM is a digital system operating at 900 MHz that uses a grid of macrocells around each country. In turn these are divided into microcells that cover areas like bus and railway stations and shopping centres, and picocells that cover individual buildings.

Example 4.1

(*a*) Calculate the audio bandwidth of a cellular radio channel. (*b*) Calculate how many r.f. channels are available in a given area.

Solution
(*a*) From equation (3.10), $25 = 2(9.5 + f_{m(max)})$
$$f_{m(max)} = 12.5 - 9.5 = 3 \text{ kHz} \quad (Ans.)$$
(*b*) Number of r.f. channels $= [(935 - 890) \times 10^6]/(25 \times 10^3)$
$$= 1800 \quad (Ans.)$$

[Note: some of these channels are needed for control and signalling purposes.]

Paging systems are operated at 138 MHz, 153 MHz (153.025 to 153.4 MHz), and 454 MHz. These systems use FSK modulation (p. 46) and a ITU−R* paging code. When called a user must go to

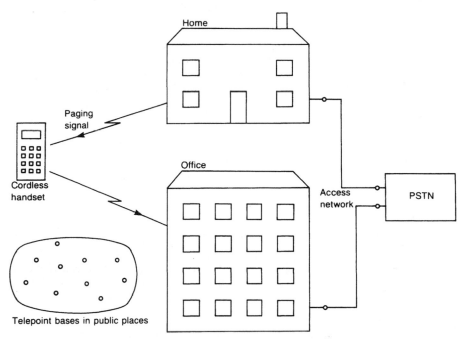

Fig. 5.1 Telepoint

a fixed position public telephone to call the office, unless the *Telepoint* system is used (see Fig. 5.1).

Cordless telephones used in the home employ a fixed base on which the handset is placed when it is not in use and a portable handset, Fig. 5.2. In the CT1 system eight channels have been allocated for use by cordless telephones; the frequencies for channel 1 are shown in the figure and the remaining frequencies are given in Table 5.4.

Personal communication networks (PCN) provide a portable telephone service to customers. The CT2 system, also known as

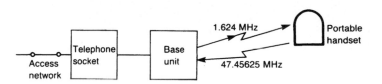

Fig. 5.2 Cordless telephone

Table 5.4 Cordless telephone (CT1) frequencies

Channel number	Base to handset (MHz)	Handset to base (MHz)
2	1.662	47.468 75
3	1.682	47.481 25
4	1.702	47.493 75
5	1.722	47.506 25
6	1.742	47.518 75
7	1.762	47.531 25
8	1.782	47.543 75

Telepoint (Fig. 5.2), allows its customers to use the same portable telephone in public places as well as at home or in the office. The user must move to be within 'line-of-sight' of a telepoint sign, when the portable telephone can be switched on and a PIN (personal identification number) entered; the caller will then be connected by a computer to the paged telephone. A large number of base stations are provided in places such as stations, garages, post offices, shopping centres and hospitals, and any telephone within about 200 metres of a Telepoint can originate a telephone call over the PSTN. CT2 operates in the frequency band 864 to 868 MHz. Another system, called DECT operating in the band 1.8 to 1.9 GHz is being introduced to give Europe-wide personal communications.

Multi-channel Telephony/Television Links

In the UHF and SHF bands, wideband radio systems, known as radio-relay systems, provide an alternative to coaxial and optical fibre cables for the provision of multi-channel telephony systems. In the UK microwave point-to-point links are provided in the frequency bands listed in Table 13.1.

Many international telephony circuits are routed over multi-channel systems which are themselves routed over a satellite communication link. The internationally agreed frequency bands for satellite systems are:

$$4/6 \text{ GHz band} \begin{cases} 5.925-6.425 \text{ GHz earth-to-satellite} \\ 3.7-4.2 \text{ GHz satellite-to-earth} \end{cases}$$

$$11/14 \text{ GHz band} \begin{cases} 14-14.5 \text{ GHz earth-to-satellite} \\ 11.45-11.7 \text{ GHz satellite-to-earth} \end{cases}$$

Data communication

For reasons that will be discussed in Chapter 9 the d.c. data signal produced by a computer or by a data terminal cannot be directly transmitted over the PSTN. Instead, the data signal must be changed into a *voice-frequency* (VF) signal using some form of digital modulation. If the signal is to be transmitted over the PSTN the available bandwidth is fairly small. At bit rates of up to 1200 bits/s FSK is employed. With FSK the data information is signalled by the transmission of either one of two different frequencies. These frequencies are 1300 Hz and either 1700 Hz or 2100 Hz. For higher bit rate transmission more complex modulation methods are necessary so that the signal can still be accommodated in the limited bandwidth offered by a telephone line.

The ITU–T and the ITU–R

To ensure compatibility between the telephone networks of different countries that play a part in a particular international telephone connection, it is necessary that carrier frequencies, bandwidths, noise levels and other parameters involved are standardized. The task of specifying the parameters of telecommunication systems destined for possible use in the international network has been given to the *International Telecommunication Union* (ITU). The ITU carries out its standardization work through three sectors: the Telecommunication Standardization Sector (ITU–T), the Radiocommunication Sector (ITU–R) and the Telecommunication Development Sector (ITU–D).

The duties of the ITU–R and the ITU–T sectors are:

(*a*) ITU–R: to study technical and operating problems relating to radiocommunications and to issue recommendations.

(*b*) ITU–T: to study technical, operating and tariff problems relating to telegraphy and telephony and to issue recommendations.

The two sectors meet at intervals to consider international telecommunication problems and policy, and include representatives from most, if not all, countries of the world. Each meeting (Plenary Assembly) produces a list of technical subjects relating to either radiocommunications or to telecommunications, the study of which should lead to improvements in the performance of systems. These questions are then referred to a number of Study Groups composed of experts from different countries who consider the problems and produce recommendations. If the recommendations are accepted by the next Plenary Assembly they are published by the ITU–T/ITU–R.

The Study Groups are listed in Tables 5.5 and 5.6.

The application of ITU–T/ITU–R recommendations to equipment that will be used for purely national routes is a matter for the particular administration concerned, and for example, not all BT's equipment conforms to recommended values. On the other hand, it is important that all equipment likely to be taken into use when an international connection is established should conform to all the relevant recommendations, otherwise many connections could have an inadequate transmission performance.

Among the ITU–T/ITU–R recommendations in current use are those covering the following topics:

(1) The synchronous digital hierarchy.
(2) Digital modulation bit rates, frequencies and phase shifts.
(3) An international trunk switching plan.
(4) The frequency stabilities required for radio transmitters operating in different frequency bands.
(5) The allocation of frequencies to different services.

Table 5.5 ITU-R Study Groups

1	Spectrum management
2	Inter-service sharing
3	Radio-wave propagation
4	Fixed-satellite services
7	Science services
8	Mobile services
9	Fixed services
10	Sound broadcasting
11	Television broadcasting

Table 5.6 ITU−T Study Groups

1	Service definitions
2	Network operation
3	Tariffs
4	Network maintenance
5	Protection against electromagnetic interference
6	Line plant
7	Data networks
8	Telematic service terminals
9	Television and sound transmission
10	Telecommunication languages
11	Switching and signalling
12	Transmission performance of networks and terminals
13	General network aspects
14	Modems and transmission techniques for data, telegraph and telematic services
15	Transmission systems and equipment

CEPT

Often the standardization procedures of the ITU−T are too slow for European countries and this delay led to the setting up of the *European Conference of Postal and Telecommunication Administrations* (CEPT). The standards work of the CEPT has since been taken over by another body, the *European Telecommunications Standards Institute* (ETSI) leaving the CEPT to deal with regulatory issues.

Exercises

5.1   Base-mobile systems are operated by a number of concerns and organizations at frequencies in the VHF and UHF bands. List some of the more important of these. State typical channel bandwidths.

5.2 State the frequency bands used for satellite communications.

5.3 State the carrier frequencies and the bandwidths for each of the following: (*a*) sound broadcasting using frequency modulation, (*b*) the sound signal of 625-line TV, and (*c*) a ship-to-shore long-distance radio-telephony link.

5.4 A taxi-cab firm uses radio to control its vehicles. State the frequency band in which the system might operate and give the probable channel bandwidths. What form of modulation would be used?

5.5 Match up the systems and the bandwidths quoted in Table 5.7.

Table 5.7

1	625-line TV broadcasts	**A**	405–525 kHz
2	Long-distance overseas sound broadcasting	**B**	25.6–26.1 MHz
		C	451–452 MHz
3	Long-distance point-to-point radio links	**D**	5.73–5.95 MHz
		E	471.25–575.25 MHz
4	Ship-to-shore telegraphy		
5	Mobile police radio		

6 Basic Transmission Line Theory

A telephone network is a combination of switching centres, such as local telephone exchanges, and transmission systems. Access to the network is gained via a local line in the access network and this employs copper conductors in a two-wire or twin line. The switching centres are interconnected by means of the trunk lines that make up the core network; these may be routed over (a) copper conductor coaxial cable, (b) optical fibre cable or (c) a microwave radio-relay system. The transmission of signals over optical fibres is considered in Chapter 7 and radio-relay systems are considered in Chapter 13.

A transmission line consists of a pair of copper conductors separated from each other by a dielectric. Two main types of line exist: the two-wire, or twin, line, shown in Fig. 6.1(a) and the coaxial line shown in 6.1(b). The two-wire line may be either open-wire or a cable pair. The coaxial pair is nearly always operated with the outer conductor earthed, since the outer then acts as an efficient screen at all operating frequencies, and is said to be *unbalanced*.

The conductors forming a pair have both *resistance* and *inductance* uniformly distributed along their length and uniformly distributed *capacitance* and *leakance* between them. These four quantities are known as the *primary coefficients* of a line.

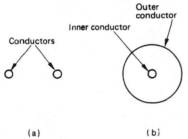

Fig. 6.1 (a) The two-wire pair, (b) the coaxial pair

The Primary Coefficients of a Line

Resistance

The resistance R of a unit length of line, or *loop* resistance, is the sum of the resistances of the two conductors comprising a pair. The unit length of a line is the kilometre.

At zero frequency the resistance of a line is the d.c. resistance R_{DC} given by

$$R_{DC} = \frac{\rho_1}{a_1} + \frac{\rho_2}{a_2} \text{ ohms per loop kilometre} \tag{6.1}$$

where ρ_1 and ρ_2 are the resistivities of the two conductors, and a_1 and a_2 are their cross-sectional areas.

At a frequency of a few kilohertz or so, a phenomenon known as *skin effect* comes into play and causes current to flow only in a thin

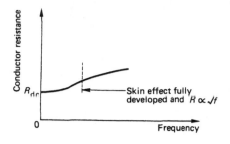

Fig. 6.2 Resistance–frequency characteristic of a transmission line

layer or 'skin' at the outer surface of the conductor. The higher the frequency the thinner this skin becomes and the smaller the cross-section of the conductor in which the current flows. Since the resistance is inversely proportional to the cross-sectional area of the 'effective' conductor, the *a.c. resistance* increases with increase in frequency. When skin effect is fully developed the a.c. resistance is proportional to the square root of the frequency, i.e.

$$R_{ac} = k_1\sqrt{f} \tag{6.2}$$

where k_1 is a constant.

Figure 6.2 shows how R_{ac} varies with frequency. Initially, little variation from the d.c. value is observed, but at higher frequencies the relationship given in equation (6.2) is true.

Inductance and Capacitance

The loop inductance L and the shunt capacitance C of a line, in henrys per loop kilometre and farads per kilometre respectively, are both more or less constant with change in frequency.

Leakance

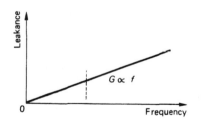

Fig. 6.3 Leakance–frequency characteristic of a transmission line

The leakance G of a line in siemens per kilometre represents the leakage of current between the conductors via the dielectric separating them, and it is very nearly the reciprocal of insulation resistance. The leakage current has two components: one component passes through the insulation between the conductors, and the other component supplies the power losses in the dielectric itself as the line capacitance is charged and discharged. Leakance increases with increase in frequency and at the higher frequencies it is directly proportional to frequency, i.e.

$$G = k_2 f \tag{6.3}$$

where k_2 is another constant. Figure 6.3 shows how the leakance of a pair of conductors varies with change in frequency.

A line can be represented by the network shown in Fig. 6.4. The line is considered to consist of a very large number of very short lengths, δl, of line connected in cascade. Each short section of line has a total shunt capacitance $C\delta l$ and shunt leakance $G\delta l$. The series inductance and resistance are really in both of the two conductors forming a pair, but it is more convenient and customary to show them all to be in the upper conductor as in Fig. 6.4. Thus, the total series inductance and resistance per elemental length δl are $L\delta l/2$ and $R\delta l/2$, respectively.

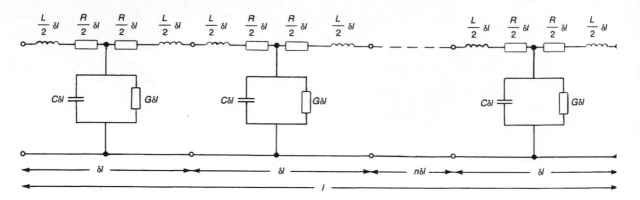

Fig. 6.4 Equivalent circuit of a line

The Secondary Coefficients of a Line

The secondary coefficients of a transmission line are its *characteristic impedance*, its *attenuation coefficient* and its *phase-change coefficient*.

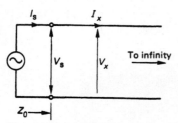

Fig. 6.5 Definition of the characteristic impedance of a line

Characteristic Impedance

The characteristic impedance Z_0 of a transmission line is the input impedance of an infinite length of that line. Figure 6.5 shows an infinite length of line; its input impedance is the ratio of the voltage V_s applied across the sending-end terminals to the current I_s flowing into the line, i.e.

$$Z_0 = V_s/I_s \text{ ohms} \tag{6.4}$$

Similarly, at any point x along the line, the ratio V_x/I_x is always equal to Z_0.

Suppose the line is now cut a finite distance from its sending-end terminals as shown in Fig. 6.6(*a*). The remainder of the line is still of infinite length and so the impedance measured at terminals 2–2′ is equal to the characteristic impedance. Thus before the line was cut, terminals 1–1′ were effectively terminated in impedance Z_0. The conditions at the input terminals will not be changed if terminals 1–1′ are closed in a physical impedance equal to Z_0, as in Fig. 6.6(*b*). This leads to a more practical definition: the characteristic impedance of a transmission line is the input impedance of a line that is itself terminated in the characteristic impedance.

A line that is terminated in its characteristic impedance is said to be *correctly terminated*.

The characteristic impedance of a line depends upon the values of the primary coefficients and also upon the frequency. At zero frequency the characteristic impedance is given by $\sqrt{(R/G)}$ ohms and then falls with increase in frequency until at higher frequencies, where $\omega L \gg R$ and $\omega C \gg G$, the impedance becomes constant at

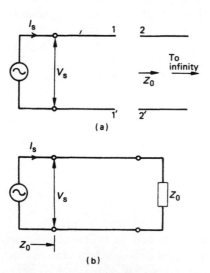

Fig. 6.6 Alternative definition of the characteristic impedance of a line

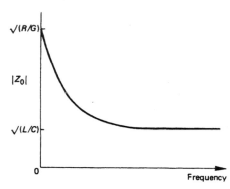

Fig. 6.7 Variation with frequency of the characteristic impedance of a line

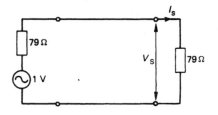

Fig. 6.8

$$Z_0 = \sqrt{(L/C)}^* \text{ ohms} \qquad (6.5)$$

This equation always applies to coaxial lines since they are only operated at frequencies high enough to make R and G negligible with respect to ωL and ωC respectively.

Figure 6.7 shows how the characteristic impedance of a line varies with frequency.

Example 6.1

A generator of e.m.f. 1 V and internal resistance 79 Ω is applied to a line having $L = 0.5$ mH/km and $C = 0.08$ μF/km. If the approximate expression, $Z_0 = \sqrt{(L/C)}$ ohms, for characteristic impedance may be assumed, calculate (a) the sending-end current, and (b) the sending-end voltage.

Solution

$$Z_0 = \sqrt{[(0.5 \times 10^{-3})/(0.08 \times 10^{-6})]} = 79\,\Omega$$

Hence, referring to Fig. 6.8,

(a) $I_s = 1/(79 + 79) = 1/158\,\text{A} \approx 6.33\,\text{mA}$ (*Ans.*)

(b) $V_s = 79 I_s = (79 \times 1)/158 = 0.5\,\text{V}$ (*Ans.*)

Most practical coaxial cables have a characteristic impedance in the range 50–75 Ω. Typical values of characteristic impedance for unloaded two-wire audio-frequency cable is 600–800 ohms at low audio frequencies, falling to about 200–300 ohms at 3000 Hz.

Attenuation Coefficient

As a current or voltage is propagated along a line its amplitude is progressively reduced or *attenuated* because of losses in the line. These losses are of two types: first, conductor losses caused by I^2R power dissipation in the series resistance, and second, dielectric losses. If the current or voltage at the sending-end terminals of the line is I_s, or V_s, then the current, or voltage, at one kilometre distance along the line is $I_1 = I_s e^{-\alpha}$, or $V_1 = V_s e^{-\alpha}$, where e is the base of the natural logarithms (2.7183) and α is the *attenuation coefficient* of the line in nepers per kilometre. In the next kilometre distance the attenuation is the same, and thus the current I_2 at the end of this distance is

$$I_2 = I_1 e^{-\alpha} = I_s e^{-\alpha} e^{-\alpha} = I_s e^{-2\alpha}$$

If the line is l kilometres long, the received current and voltage are given, respectively, by

$$I_r = I_s e^{-\alpha l} \qquad (6.6)$$

$$V_r = V_s e^{-\alpha l} \qquad (6.7)$$

*The full expression for the characteristic impedance of a line is

$$Z_0 = \sqrt{\left(\frac{R + j\omega L}{G + j\omega C}\right)} \text{ ohms}$$

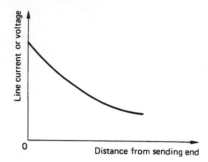

Fig. 6.9 Decay of current and voltage along a transmission line

$$I_x = I_s e^{-\alpha x} \quad \text{or} \quad V_x = V_s e^{-\alpha x}$$

Thus both current and voltage waves decay exponentially as they are propagated along the line as shown by Fig. 6.9.

Example 6.2

A correctly terminated line has a characteristic impedance of 800 Ω, an attenuation coefficient of 0.3 Np and it is 4 km in length. Calculate the current in the load when the voltage applied across the sending terminals of the line is 2 V.

Solution

From equation (6.7)

$$V_r = 2e^{-0.3 \times 4} = 2e^{-1.2}$$
$$= 0.602 \text{ V}$$
$$I_r = V_r/Z_0 = 0.602/800 = 0.753 \text{ mA} \quad (Ans.)$$

Alternatively, the attenuation coefficient of a line may be quoted in decibels per kilometre. 1 neper = 8.686 dB

Example 6.3

Repeat Example 6.2 using decibels.

Solution

0.3 Np = 0.3 × 8.686 dB = 2.6058 dB and so the total line loss is

$$2.6058 \times 4 = 10.42 \text{ dB}$$

Therefore, $10.42 = 20 \log_{10}(2/V_r)$

or $10.42/20 = 0.52 = \log_{10}(2/V_r)$

$$10^{0.52} = 3.31 = 2/V_r \quad \text{or} \quad V_r = 0.604 \text{ V}$$

$$I_r = 0.604/800 = 0.76 \text{ mA} \quad (Ans.)$$

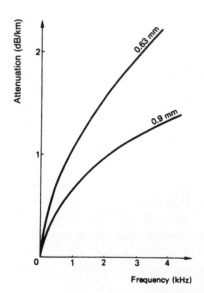

Fig. 6.10 Attenuation–frequency characteristics of audio-frequency star-quad cable

The attenuation of a cable is proportional to the length of the line. Thus if α is 3 dB/km at a particular frequency, the overall loss of a 2 km length of line is 6 dB, of a 4 km length of line is 12 dB, and so on.

Both the resistance and the conductance of a line increase with increase in frequency, and since these are the power-dissipating components the attenuation of the line increases with increase in frequency. Figure 6.10 shows how the attenuation coefficient of two types of cable varies with frequency. The attenuation continues to increase at frequencies above those shown in the figure and may be as large as 25 dB/km at 1 MHz. Because of excessive losses like this wideband telecommunication systems, which employ frequencies o

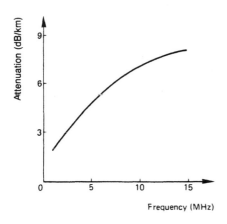

Fig. 6.11 Attenuation—frequency characteristics of a coaxial pair

some hundreds of kilohertz upwards, use coaxial cable as the transmission medium. The attenuation—frequency characteristics of a coaxial pair are shown in Fig. 6.11. The coaxial pair cannot be used at frequencies below about 60 kHz because the earthed outer conductor would then cease to act as an efficient screen.

Example 6.4

A 6 km length of 0.63 mm star-quad cable is used as a data link. Determine the ratio of the attenuations experienced by the fundamental and the third and fifth harmonics of the data waveform when the baud speed is (a) 1200, (b) 2400.

Solution

(a) At 1200 bauds the maximum fundamental frequency is 600 Hz, the third harmonic is 1800 Hz, and the fifth harmonic is 3000 Hz. From Fig. 6.10,

at   600 Hz $\alpha \simeq 0.8$ dB/km
at 1800 Hz $\alpha \simeq 1.4$ dB/km
at 3000 Hz $\alpha \simeq 1.9$ dB/km

Thus the loss of a 6 km length of this cable is 4.8 dB at 600 Hz, 8.4 dB at 1800 Hz, and 11.4 dB at 3000 Hz.

4.8 dB is a voltage ratio of 1.74, 8.4 dB is a voltage ratio of 2.63, and 11.4 dB is a voltage ratio of 3.72. Therefore

Ratio of attenuation = 2.63/1.74 = 1.5   (third)
Ratio of attenuation = 3.72/1.74 = 2.14   (fifth)   (*Ans.*)

(b) At 2400 bauds the maximum fundamental frequency is 1200 Hz, the third harmonic is 3600 Hz, and the fifth harmonic is 6000 Hz. From Fig. 6.10,

at 1200 Hz $\alpha = 1.1$ dB/km
at 3600 Hz $\alpha = 3.0$ dB/km
at 6000 Hz $\alpha = 4.5$ dB/km

Hence the loss of a 6 km length of the cable at each frequency is 6.6 dB, 18 dB and 27 dB respectively.

The attenuation ratios are:

fundamental/third harmonic: 7.94/2.14 = 3.7, and
fundamental/fifth harmonic: 22.39/2.14 = 10.5   (*Ans.*)

Cables Connected in Cascade

When cables are connected in cascade their attenuation—frequency characteristics are *additive* as is shown by the following example.

Example 6.5

The attenuation—frequency characteristics of four types of audio cable are given by Table 6.1.

Table 6.1

	Attenuation (dB/km)		
	800 Hz	*1600 Hz*	*2000 Hz*
Cable A	1.26	1.80	1.95
Cable B	1.13	1.61	1.74
Cable C	0.87	1.23	1.37
Cable D	0.59	0.81	0.89

Plot the overall attenuation—frequency characteristics of links consisting of the cascade connection of

(i) 2 km of cable A, 1 km of cable B, and 2 km of cable C,
(ii) 3 km of cable A, 2 km of cable C, 1.5 km of cable D, and 2.5 km of cable C.

Assume that there are no reflections at any of the cable junctions.

Solution
The required graphs are shown plotted in Fig. 6.12 and are obtained from the data given in Table 6.2.

Table 6.2

	Attenuation (dB)		
	800 Hz	*1600 Hz*	*2000 Hz*
Link (i)			
2 km of cable A	2.52	3.60	3.90
1 km of cable B	1.13	1.61	1.74
2 km of cable C	1.74	2.46	2.74
Total loss	5.39	7.67	8.38
Link (ii)			
3 km of cable A	3.78	5.40	5.85
2 km of cable C	1.74	2.46	2.74
1.5 km of cable D	0.89	1.22	3.43
2.5 km of cable C	2.18	3.08	3.43
Total loss	8.59	12.16	13.36

High-frequency Attenuation

At frequencies where $\omega L \gg R$ and $\omega C \gg G$ the attenuation coefficient is given by

$$\alpha = R/2Z_0 + GZ_0/2 \text{ nepers per kilometre} \qquad (6.8)$$

Usually, the second term is negligibly small and

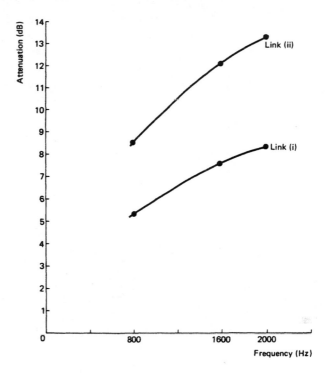

Fig. 6.12

$$\alpha = R/2Z_0 \qquad\qquad (6.9)$$

Example 6.6

A coaxial cable has a loss of 3.5 dB/km at 1 MHz. Calculate its loss at 4 MHz if (*a*) the dielectric loss is negligible, and (*b*) the dielectric loss is 10% of the total.

Solution

(*a*) Loss at 4 MHz = loss at 1 MHz $\times \sqrt{[(4 \times 10^6)/(1 \times 10^6)]}$

$$= 3.5 \times 2 = 7 \text{ dB/km} \quad (Ans.)$$

(*b*) Dielectric loss at 1 MHz = 0.35 dB/km
Conductor loss at 1 MHz = 3.15 dB/km

Loss at 4 MHz = $3.15 \times \sqrt{(4/1)} + 0.35 \times 4/1$

$$= 6.3 + 1.4 = 7.7 \text{ dB/km} \quad (Ans.)$$

Phase-change Coefficient

A current or voltage wave travels along a line with a finite velocity and so the current, or voltage, at the end of a metre length of line lags the current, or voltage, entering that length. The phase difference between the line currents, or voltages, at two points which are one metre apart is known as the *phase-change coefficient* β of the line. β is measured in radians per kilometre. In each kilometre distance of a line the same phase shift is introduced; consequently for a line l kilometres in length the received current will lag the sending-end current by βl radians.

At high frequencies where $\omega L \gg R$ *and* $\omega C \gg G$

$$\beta = \omega \sqrt{(LC)} \text{ rad/km} \tag{6.10}$$

Example 6.7

A correctly terminated transmission line has $Z_0 = 500 \, \Omega$, $\alpha = 1 \text{ dB km}^{-1}$ and $\beta = 30° \text{ km}^{-1}$, and it is 3 km long. A 500 Ω source, of e.m.f. 2 V, is applied to the sending-end terminals of the line. Calculate (*a*) the magnitude of the received current, and (*b*) its phase relative to the sending-end voltage.

Solution

(*a*) Since the line is correctly terminated its input impedance is equal to its characteristic impedance of 500 Ω. Therefore

$$I_s \quad = 2/(500 + 500) = 2 \text{ mA}$$

Line loss $= 3 \times 1 = 3 \text{ dB}$

The load and input impedances of the line are both 500 Ω. Hence,

$$3 = 20 \log_{10}(2/I_r)$$
$$10^{0.15} = 2/I_r \simeq 1.414 \quad \text{or} \quad I_r = 1.414 \text{ mA} \quad (Ans.)$$

(*b*) The phase shift introduced by the line is $3 \times 30° = 90°$; therefore

I_r lags V_s by $90°$ (*Ans.*)

Velocity of Propagation

Phase Velocity of Propagation

The phase velocity v_p of a line is the velocity with which a sinusoidal wave travels along that line. Any sinusoidal wave travels with a velocity of one wavelength per cycle. There are f cycles per second and so a wave travels with a velocity of λf metres per second, i.e.

$$v_p = \lambda f \text{ m/s} \tag{6.11}$$

where λ is the wavelength and f is the frequency of the sinusoidal wave.

In one wavelength a phase change of 2π radians occurs, and hence the phase change per metre is $2\pi/\lambda$ radians, and this is also equal to the phase-change coefficient. Thus

$$\beta = 2\pi/\lambda \quad \text{or} \quad \lambda = 2\pi/\beta \tag{6.12}$$

and

$$v_p = 2\pi/\beta \times f = \omega/\beta \text{ m/s} \qquad (6.13)$$

Example 6.8

A correctly terminated line has an attenuation of 3 dB/km and a phase-change coefficient of 0.1 rad/km at 1500 Hz. When 1 V is applied across its sending-end terminals the voltage across the load is 125 mV. Determine the time it takes for a 1500 Hz signal to travel over the line.

Solution
Attenuation $= 20 \log_{10}(1/0.125) = 18$ dB
Line length $= 18/3 = 6$ km
$v_p = (2\pi \times 1500)/0.1 = 9.425 \times 10^4$ km/s
$t = 1/v_p = 6/(9.425 \times 10^4) = 63.66 \,\mu s$ (*Ans.*)

Example 6.9

A 0.5 m length of loss-free coaxial line is used for VHF measurements. The line has a capacitance of 88 pF/m and an inductance of 220 nH/m. Calculate (*a*) its characteristic impedance, (*b*) its velocity of propagation as a percentage of the velocity of light and (*c*) the time taken for a wave at (i) 30 MHz and (ii) 300 MHz to travel over the line.

Solution
(*a*) $Z_0 = \sqrt{[(220 \times 10^{-9})/(88 \times 10^{-12})]} = 50 \,\Omega$ (*Ans.*)

(*b*) $v_p = \omega/\beta = \omega/\omega\sqrt{(LC)} = 1/\sqrt{(LC)}$
$= 1/\sqrt{[(88 \times 10^{-12})(220 \times 10^{-9})]} = 2.273 \times 10^8$ m/s
$= 2.273/3 = 75.7\%$ of velocity of light (*Ans.*)

(*c*) At VHF the phase velocity $v_p = 1/\sqrt{(LC)}$ and it is independent of frequency.
(i) $t = d/v_p = 0.5/(2.273 \times 10^8) = 2.2$ ns (*Ans.*)
(ii) $t = 2.2$ ns (*Ans.*)

Group Velocity and Group Delay

Any repetitive, non-sinusoidal waveform contains components at a number of different frequencies, each of which will be propagated along a transmission line with a phase velocity given by Equation (6.13). For all these components to travel with the same velocity and arrive at the far end of the line at the same moment, it is necessary for the phase change coefficient β of the line to be a linear function of frequency, i.e. for ω/β to be a constant at all frequencies. It is only at radio frequencies that practical lines satisfy this requirement. At lower frequencies β varies with frequency in a non-linear manner. Figure 6.13 shows the phase change coefficient—frequency characteristic of a typical audio-frequency cable.

Suppose a signal consisting of a 1000 Hz fundamental plus 50% third harmonic is applied to a 10 km length of a pair in this cable.

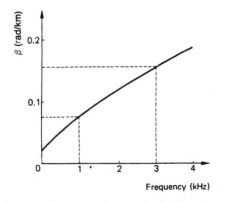

Fig. 6.13 Phase change coefficient—frequency characteristic of a transmission line

The fundamental frequency component will be propagated with a phase velocity

$$v_p = \omega/\beta = (2\pi \times 1000)/(0.075 \times 10^{-3})$$

or

$$83.78 \times 10^6 \, \text{m/s}$$

and the component at the third harmonic will propagate with a phase velocity of

$$\omega/\beta = (6\pi \times 1000)/(0.155 \times 10^{-3}) \quad \text{or} \quad 121.61 \times 10^6 \, \text{m/s}$$

This means that the harmonic component will arrive at the far end of the line t seconds before the fundamental arrives.

$$\text{Time } t = (\text{length of line})/(\text{velocity difference})$$
$$= 10^4/[(121.61 - 83.78) \times 10^6] = 0.264 \, \text{ms}$$

This time is 0.264 times the periodic time of the fundamental frequency component and so the third harmonic component leads the fundamental component by an angle of $0.264 \times 360°$ or approximately $95°$. Figures 6.14(a) and (b) show, respectively, the resultant waveforms at the beginning and at the end of the line; it is evident that waveform distortion has taken place.

It is customary to consider the *group velocity* of a complex wave rather than the phase velocities of its individual frequency components. Group velocity is the velocity with which the *envelope* of the resultant waveform is propagated. Figure 6.15, for example, illustrates the meaning of the term group velocity when it is applied to the transmission of an amplitude-modulated wave over a line. The envelope travels at the group velocity, while the carrier, which is one of the component frequencies of the modulated wave, propagates with its particular phase velocity.

If a narrow band $\omega_2 - \omega_1$ of frequencies is transmitted over a line and at these two frequencies the phase-change coefficients of the line are β_2 and β_1 respectively, then the group velocity V_g is given by equation (6.14), i.e.

$$V_g = (\omega_2 - \omega_1)/(\beta_2 - \beta_1) \, \text{m/s} \tag{6.14}$$

The *group delay* of a line is the product of the length of the line and reciprocal of its group velocity. Hence, it is the propagation time of a complex, i.e. non-sinusoidal, signal.

Example 6.10

A 1200 baud data signal is transmitted over a line whose phase-change coefficient characteristic is shown plotted in Fig. 6.13. The bandwidth of the transmitted signal is limited so that only frequencies up to the fifth harmonic are included. Determine (a) the group velocity of the signal, (b) the group delay if the line is 3 km in length.

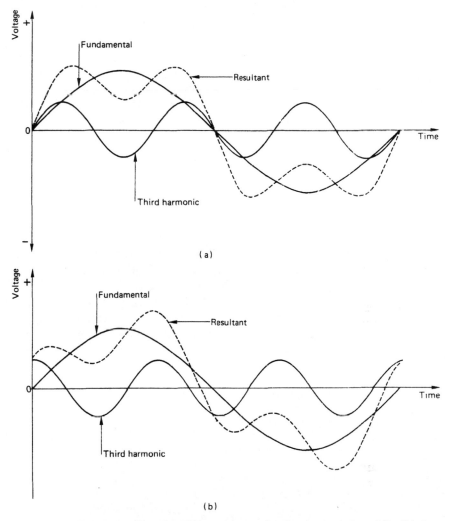

Fig. 6.14 Showing the envelope of a fundamental and its third harmonic (*a*) at the beginning and (*b*) at the end of a line in which the ratio ω/β is not constant

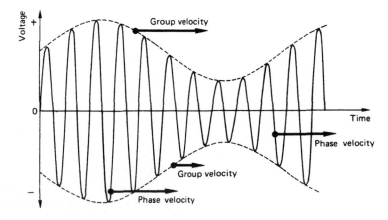

Fig. 6.15 Group and phase velocities of an amplitude-modulated wave

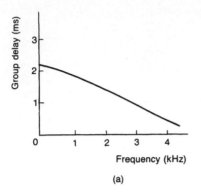

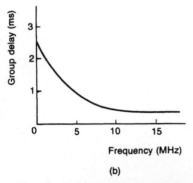

Fig. 6.16 Group delay–frequency characteristics of (a) an audio-frequency cable, (b) a coaxial cable

Solution

The maximum fundamental frequency of a 1200 baud data waveform is 600 Hz. The third and the fifth harmonics are 1800 and 3000 Hz. Thus the lowest and the highest frequencies to be transmitted are 600 Hz and 3000 Hz respectively. From the graph, the values of the phase-change coefficient at these two frequencies are 0.056×10^{-3} rad/km and 0.155×10^{-3} rad/km respectively.

$$V_g = \frac{2\pi(3000 - 600)}{0.155 - 0.056} = 152 \times 10^3 \text{ km/s}$$

The group delay is

$$[1/(152 \times 10^3)]3 = 19.7 \,\mu s$$

The group delay of a cable is proportional to the length of that cable. Thus if a cable has a group delay of 8 μs per kilometre at a particular frequency the group delay of a 10 km length will be 80 μs.

Two typical group delay–frequency characteristics are given in Fig. 6.16. Figure 6.16(a) shows clearly that the group delay on an audio-frequency cable decreases with increase in frequency; obviously it will also increase as the length of the line is increased. When the group delay of a circuit varies with frequency, the different components of a complex signal will not arrive at the end of the line at the same time and *group delay distortion* will be present. Because the group delay of a line is to some extent determined by its length, *relative group delay* is often quoted instead. The relative group delay is the difference between the propagation delays at a given frequency and those at a chosen reference frequency.

The Effect of Cables on Analogue and Digital Signals

For a reproduced sound signal to appear as natural as possible within the necessary bandwidth restrictions, it is essential that the original amplitude relationships between the fundamental component and the various harmonics are retained. As a signal is propagated along a transmission line it will be attenuated, and this attenuation is greater at the higher frequencies than at the lower. The effect of line attenuation is, therefore, to reduce the amplitudes of the harmonics relative to the fundamental and to make the received sound seem unnatural. The effect on short telephone lines is slight and can be tolerated but longer telephone circuits and all music circuits are normally *equalized* to overcome this problem.

With data and television transmissions the signal waveform must be retained and this means that the various component frequencies must keep both their amplitude and phase relationships relative both to one another and to the fundamental component. Thus, both amplitude–frequency and group delay–frequency distortion must be small. The bandwidth occupied by a television signal is several megahertz and consequently both amplitude and group delay–

frequency distortion are of importance. Lines for the transmission of television signals are generally fitted with both attenuation and group delay equalizers. Modems often include an adaptive equalizer.

Mismatched Transmission Lines

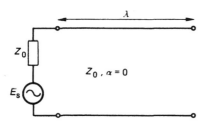

Fig. 6.17 Open-circuited loss-free line

Very often a transmission line is operated with a terminating impedance that is not equal to the characteristic impedance of the line. When the impedance terminating a line is not equal to the characteristic impedance, the line is said to be *incorrectly terminated* or *mismatched*. Since the line is not matched, the load is not able to absorb all the incident power and so some of the power is *reflected* back towards the sending-end of the line.

Figure 6.17 shows a loss-free line whose output terminals are open circuited. The line has an electrical length of one wavelength and its input terminals are connected to a generator of e.m.f. E_s volts and internal resistance Z_0 ohms.

When the generator is first connected to the line, the input impedance of the line is equal to its characteristic impedance Z_0. An *incident* current of $E_s/2Z_0$ then flows into the line and an *incident* voltage of $E_s/2$ appears across the input terminals. These are, of course, the same values of sending-end current and voltage that flow into a correctly terminated line. The incident current and voltage waves propagate along the line, being phase shifted as they travel. Since the electrical length of the line is one wavelength, the overall phase shift experienced is 360°.

Since the output terminals of the line are open circuited, no current can flow between them. This means that *all* of the incident current must be *reflected* at the open circuit. The total current at the open circuit is the phasor sum of the incident and reflected currents, and since this must be zero the current must be reflected with 180° phase shift. The incident voltage is also totally reflected at the open circuit but with zero phase shift. The total voltage across the open-circuited terminals is twice the voltage that would exist if the line were correctly terminated. The reflected current and voltage waves propagate along the line towards its sending end, being phase shifted as they go. When the reflected waves reach the sending end, they are completely absorbed by the impedance of the matched source.

At any point along the line, the total current and voltage is the phasor sum of the incident and reflected currents and voltages. The way in

Fig. 6.18 The r.m.s. value of the total current along a loss-free open-circuited line

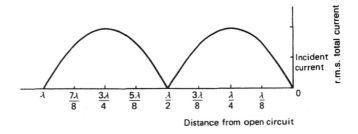

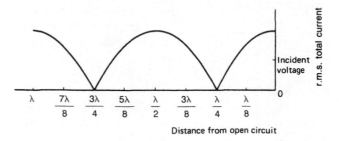

Fig. 6.19 The r.m.s. value of the total voltage along a loss-free open-circuited line

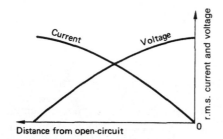

Fig. 6.20 Standing waves of current and voltage on a λ/4 open-circuited loss-free line

which the r.m.s. line current varies with distance from the open circuit is shown by Fig. 6.18. The points at which maxima (*antinodes*) and minima (*nodes*) of current occur are always the same and do not vary with time. Because of this the waveform of Fig. 6.18 is said to be a *standing wave*.

At the open-circuited output terminals, the incident and reflected voltages are in phase and the total voltage is twice the incident voltage. Hence the total voltage at any point along the line is the phasor sum of the incident and reflected voltages and its r.m.s. value varies in the manner shown in Fig. 6.19.

Two things should be noted from Fig. 6.18 and Fig. 6.19. First, the *voltage standing-wave* pattern is displaced by λ/4 from the *current standing-wave* pattern, i.e. a current antinode occurs at the same point as a voltage node and vice versa. Second, the current and voltage values at the open circuit are repeated at λ/2 intervals along the length of the line; this remains true for any longer length of loss-free line.

When the output terminals of a loss-free line are short circuited, the conditions at the termination are reversed. There can be no voltage across the output terminals but the current flowing is twice the current that would flow in a matched load. This means that at the short circuit the incident current is totally reflected with zero phase shift and the incident voltage is totally reflected with 180° phase shift. Thus, Fig. 6.18 shows how the r.m.s. voltage on a short-circuited line varies with distance from the load, and Fig. 6.19 shows how the r.m.s. current varies along a short-circuited line. The standing waves of current and voltage on a λ/4 length of open-circuited line are easily obtained and are shown by Fig. 6.20.

Standing Wave Ratio

An important parameter of any mismatched transmission line is its *voltage standing-wave ratio* or VSWR. The VSWR is the ratio of the maximum voltage on the line to the minimum voltage, i.e.

$$S = V_{max}/V_{min} \qquad (6.15)$$

The presence of a standing wave on a mismatched line is undesirable for several reasons and very often measures are taken to approach the matched condition and hence to minimize reflections. The reasons why standing waves on a line should be avoided if at all possible are as follows.

(a) Maximum power is transferred from a transmission line to its load when the load impedance is equal to the characteristic impedance. When a load mismatch exists, some of the incident power is reflected at the load and the transfer efficiency is reduced.

(b) The power reflected by a mismatched load will propagate, in the form of current and voltage waves, along the line towards its sending end. The waves will be attenuated as they travel and so the total line loss is increased.

(c) At a point of voltage maximum the line voltage may be anything up to twice as great as the incident voltage. For low-power lines the increased voltage will not matter. For a line connecting a high power radio transmitter to an aerial, for example, the situation is quite different. Care must be taken to ensure that the maximum line voltage will not approach the breakdown voltage of the line's insulation. This means that for any given value of VSWR there is a corresponding peak value for the incident voltage and hence for the maximum power that the line is able to transmit. A high VSWR on a line can result in dangerously high voltages appearing at the voltage antinodes. Great care must then be taken by maintenance staff who are required to work on, or near to, the line system.

Measurement of VSWR

The VSWR on a mismatched transmission line can be determined by measuring the maximum and minimum voltages that are present on the line. In practice, the measurement is generally carried out using an instrument known as a *standing-wave indicator*. Measurement of VSWR not only shows up the presence of reflections on a line but it also offers a most convenient method of determining the nature of the load impedance.

The measurement procedure is as follows. The VSWR is measured, using a detector, a galvanometer, or a VSWR meter, and the distance in wavelengths from the load to the voltage minimum nearest to the load is determined. The values obtained allow the magnitude and angle of the load impedance to be determined, usually with the assistance of a graphical aid known as a Smith Chart.

Example 6.11

A loss-free line has its VSWR measured at a frequency of 1 GHz. The measured values are $V_{max} = 12$ mV and $V_{min} = 3$ mV and the first voltage minimum is 5 cm from the load. Calculate (a) the wavelength on the line, (b) the VSWR, (c) the distance in wavelengths between the load and the first voltage minimum, (*d*) the distance in wavelengths between the load and the first voltage maximum, and (*e*) the distance in centimetres between two adjacent voltage maxima, or minima. Sketch the waveform of the voltage on the line.

Solution

(*a*) $\lambda = (3 \times 10^8)/(1 \times 10^9) = 0.3 \, \text{m} = 30 \, \text{cm}$ (*Ans.*)

(*b*) VSWR $= 12/3 = 4$ (*Ans.*)

(*c*) $5 \, \text{cm} = \lambda/6$ (*Ans.*)

(*d*) $\lambda/6 + \lambda/4 = 5\lambda/12$ (*Ans.*)

(*e*) 15 cm   (*Ans.*)

The waveform is shown in Fig. 6.21.

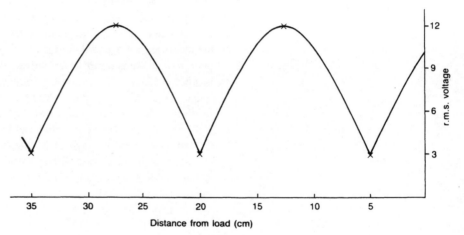

Fig. 6.21

Distance from load (cm)

Methods of Matching

A quarter-wavelength ($\lambda/4$) length of low-loss line possesses an important impedance-transforming property which is used for a wide variety of purposes at the higher frequencies.

A load impedance Z_L can be transformed into any desired value of input impedance Z_{in} by the suitable choice of the characteristic impedance Z_0 of a $\lambda/4$ length of low-loss line. The required value of the characteristic impedance is given by

$$Z_0 = \sqrt{(Z_L Z_{in})} \qquad (6.16)$$

One common application of the *quarter-wave* ($\lambda/4$) *transformer* is the matching of a transmission line to a load impedance which is not equal to Z_0. Figure 6.22(a) shows a 600 Ω transmission line which is to be connected to a 300 Ω load impedance. If the line is directly connected to the load, reflections will occur and a voltage standing-wave pattern will appear on the line. To avoid this happening, the load impedance must be transformed into 600 Ω so that a matched system is obtained. A $\lambda/4$ matching section should be connected

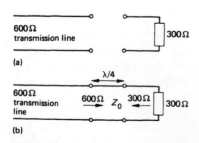

Fig. 6.22 (*a*) 600 Ω line and mismatched 300 Ω load, (*b*) use of a $\lambda/4$ section of low-loss line to obtain a matched system

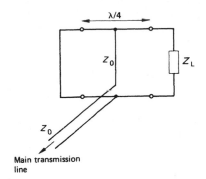

Fig. 6.23 Use of a λ/4 matching stub

between the end of the 600 Ω line and the 300 Ω load as shown by Fig. 6.22(b). For the input impedance of the λ/4 section to be equal to 600 Ω, its characteristic impedance must be

$$Z_0 = \sqrt{(600 \times 300)} = 424.3 \, \Omega$$

Another method of using a λ/4 length of low-loss line as a matching device is shown by Fig. 6.23. The λ/4 section has one pair of terminals short-circuited and its other pair connected across the load impedance. The impedance of the λ/4 line will vary along its length from Z_L to zero and at some point it will be equal to the characteristic impedance of the main line. If the main line is connected to the λ/4 section at this point, it will be effectively correctly terminated and there will be no reflections.

Example 6.12

A resistance of 200 Ω is to be matched to a 600 Ω line by a λ/4 transformer. Determine the required characteristic impedance of the matching section.

Solution
$$Z_0 = \sqrt{(200 \times 600)} = 346 \, \Omega \qquad (Ans.)$$

Exercises

6.1 A coaxial pair has a loss of 10 dB at a frequency of 1 MHz. Calculate its loss at 4 MHz, assuming that the dielectric losses are negligibly small.

6.2 A coaxial cable has a loss of 4 dB/km at 2 MHz. If dielectric losses are negligible, what is the loss of 2.5 km length of this cable at 4 MHz?

6.3 A generator of e.m.f. 50 V and internal impedance 600 Ω is applied to the sending-end terminals of a line having a characteristic impedance of 600 Ω and an attenuation coefficient of 1 dB/km. Calculate the current that flows in the correctly terminated load resistance if the line is 20 km long.

6.4 A correctly terminated line has $Z_0 = 600 \, \Omega$ and $\alpha = 0.8$ dB km^{-1}. It is fed by a generator of e.m.f. 10 V and internal resistance 600 Ω. If the power dissipated in the load is 5 mW calculate the length of the line.

6.5 Explain the difference between the phase velocity and the group velocity of a line. What is meant by group delay and by group delay–frequency distortion? Explain the effects that group delay–frequency distortion may have upon (*a*) an analogue signal, and (*b*) a digital signal.

6.6 What is meant by the term characteristic impedance when it is applied to a transmission line? A line has $Z_0 = 75 \, \Omega$ and is correctly terminated. Determine its input impedance.

6.7 A signal consisting of a 5 kHz fundamental and its third harmonic is transmitted over a transmission line. The phase-change coefficient of the line is 10°/km at 5 kHz and 24°/km at 15 kHz. Calculate the group velocity of the signal.

6.8 A current of 12 mA flows into a 4 km length of transmission line. The current that flows in the correctly terminated load is 4 mA. Calculate the attenuation coefficient of the line.

6.9 A cable has an attenuation coefficient of 1 dB/km at 1000 Hz and 1.8 dB/km at 3000 Hz. What will be the loss at each frequency of a 6 km length of this line?

The signal applied across the sending-end terminals of this line consists of a 10 V, 1000 Hz fundamental plus 25% third harmonic. Calculate the voltage at each frequency across the correctly terminated far end of the line.

6.10 A loss-free line has its far-end terminals connected to an impedance that is not equal to Z_0. Explain why standing waves will appear on the line. If the maximum and minimum voltages along the line are 20 V and 4 V respectively, calculate the VSWR.

7 Optical Fibre Systems

Visible and infrared light extends over a range of wavelengths from about $0.4\,\mu$m to about 1 mm but the use of fibre optics is, at present, restricted to the approximate waveband of $0.8\,\mu$m to $1.6\,\mu$m. The energy carried by light waves in this waveband can be transmitted by glass fibres which act as *dielectric waveguides*. Essentially, an optical fibre consists of a cylindrical glass *core* that is surrounded by a glass *cladding*. The use of an optical fibre to transmit light energy offers a number of advantages over the more conventional transmission techniques using copper conductors. These advantages are as follows.

(a) Light-weight, small-dimensioned cables so smaller ducts are necessary.
(b) Very wide bandwidth of about 50 000 GHz.
(c) Freedom from electromagnetic interference.
(d) Low attenuation of less than 0.4 dB/km.
(e) High reliability and long life.
(f) Cheap raw materials.
(g) Negligible crosstalk between fibres in the same cable.
(h) Fewer repeaters and power feed equipments are required.

Optical fibre technology is now well established as an alternative to copper cable for use in long-distance, wideband communication systems. Optical fibre is particularly suited to the transmission of digital signals, and the most important telecommunications applications are in connection with PCM multi-channel telephone systems. In 1993 optical fibre accounted for approximately 85% of the trunk network of the UK.

Reflection and Refraction

When a light wave travelling in one medium passes into another medium, its direction of travel will probably be changed. The light wave is said to be *refracted*. The ratio

$$\frac{\text{sine of angle of incidence}}{\text{sine of angle of refraction}} = \frac{\text{sine } \varphi_i}{\text{sine } \varphi_r} = \eta \qquad (7.1)$$

is a constant for a given pair of media. This constant is known as the *refractive index* η for the two media. If one of the two media is

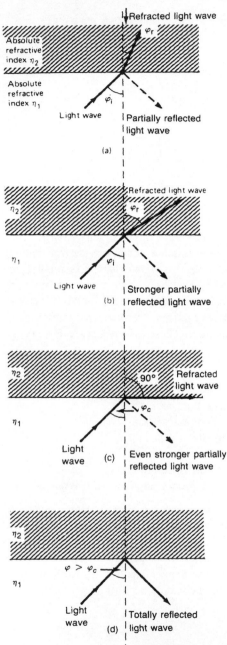

Fig. 7.1 Reflection and refraction at the boundary of two media with differing refractive indices: (a) $\eta_2 > \eta_1$, (b) $\eta_1 > \eta_2$, (c) critical angle of incidence, (d) total reflection

air, whose refractive index is unity, the *absolute refractive index* of the other medium is obtained. If the absolute refractive indices of the two media are η_1 and η_2 respectively, then

$$\sin \varphi_i / \sin \varphi_r = \eta_2 / \eta_1 \qquad (7.2)$$

When the light passes from a medium of lower refractive index into a medium of higher refractive index, i.e. when $\eta_2 > \eta_1$, the wave will be bent towards the normal (Fig. 7.1(a)). Conversely, if $\eta_1 > \eta_2$, the light wave will be refracted away from the normal as shown by Fig. 7.1(b). Some of the incident light energy will be reflected at the boundary of the two media. As the angle of incidence is increased, the angle of refraction is also increased and the partially reflected wave gets stronger.

Eventually the point is reached at which the angle of refraction φ_r is 90°. Then $\sin \varphi_r = 1$ and the refracted wave will travel parallel to the boundary of the two media as shown by Fig. 7.1(c). If the angle of incidence is still further increased the light wave will be *totally reflected* and then there is no longer a refracted wave. The angle of incidence at which total reflection first occurs is known as the *critical angle* φ_c and it is shown in Fig. 7.1(d). The value of φ_c depends on the absolute refractive indices of the two media.

When $\varphi_i = \varphi_c$. $\sin \varphi_i / \sin \varphi_r = \sin \varphi_c / 1 = \eta_2 / \eta_1$, or

$$\varphi_c = \sin^{-1}(\eta_2 / \eta_1) \qquad (7.3)$$

The angle of reflection is always equal to the angle of incidence.

Example 7.1

Light is passed through glass of refractive index 1.49. Calculate the angle of incidence at which total internal reflection will take place.

Solution
For total internal reflection $\varphi_r = 90°$ and so $\sin \varphi_r = 1$.
  Therefore,

$$1.49 = 1/\sin \varphi_i.$$

$$\sin \varphi_i = 1/1.49 = 0.67$$

$$\varphi_i = \sin^{-1} 0.67 = 42.1° \quad (Ans.)$$

Suppose that now another medium, also of refractive index η_2, is placed on the other side of the lower η_1 medium, as shown by Fig. 7.2. Provided the angle of incidence φ_i of the input light wave is larger than the critical value φ_c, the light wave will be able to propagate along the inner medium by means of a series of total reflections. Any other light waves that are also incident on the upper boundary at an angle $\varphi_i > \varphi_c$ will also propagate along the inner medium. Conversely, any light wave that is incident upon the upper boundary with an angle of incidence less than the critical value, i.e.

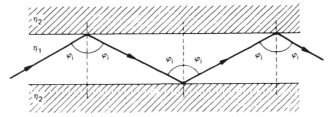

Fig. 7.2 Light wave propagating by multiple reflections

$\varphi_i < \varphi_c$, will pass into the upper medium and there be lost by *scattering* and/or *absorption*.

In an optical fibre the middle medium of refractive index η_1 is known as the *core* and the outer medium of refractive index η_2 is known as the *cladding*.

The Propagation of Light in an Optical Fibre

Figure 7.3 shows the arrangement of an optical fibre. A glass core is surrounded by a glass cladding of slightly smaller refractive index. Typical values for the refractive indices of the core and the cladding are 1.5 and 1.49 respectively. If a light wave is incident upon the core—cladding interface at an angle of incidence which is greater than the critical value φ_c, the wave will propagate along the fibre as shown by Fig. 7.2. For a light wave to be incident upon the core—cladding interface at the critical angle it must enter the fibre at the maximum *acceptance angle* $\theta_{A(max)}$, see Fig. 7.4. For acceptance angles smaller than $\theta_{A(max)}$ the angle of incidence on the core—cladding interface will be larger than the critical value. There is thus a range of acceptance angles, measured from the longitudinal axis of the fibre, over which a light wave may be incident upon the core and be transmitted along the fibre. These acceptance angles are all included within the *cone of acceptance* shown in Fig. 7.5.

$$
\begin{aligned}
\eta_{air}\sin \theta_{A(max)} &= \eta_1\sin(90° - \varphi_c) = \eta_1 \cos \varphi_c \\
\sin^2\theta_{A(max)} &= \eta_1^2 \cos^2\varphi_c \\
&= \eta_1^2(1 - \sin^2\varphi_c) \\
&= \eta_1^2[1 - (\eta_2/\eta_1)^2] \\
\sin \theta_{A(max)} &= \sqrt{(\eta_1^2 - \eta_2^2)}
\end{aligned}
$$

(7.4)

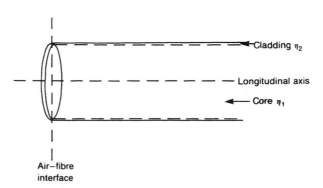

Fig. 7.3 Optical fibre

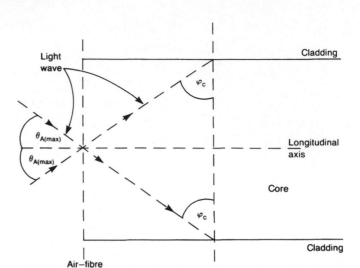

Fig. 7.4 Acceptance angle

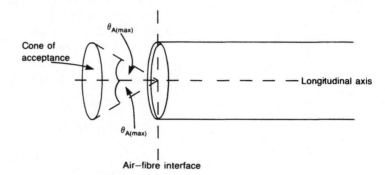

Fig. 7.5 Cone of acceptance

Figure 7.6 shows what happens when a light wave enters an optical fibre at (a) an angle less than angle $\theta_{A(max)}$, (b) an angle $\theta_{A(max)}$, and (c) an angle that is greater than angle $\theta_{A(max)}$. In case (a) the light wave travels down the fibre by a series of reflections at the core–cladding interface, and in case (b) the light wave travels along the interface and there are also some refracted waves. In the third case there is no propagation of the light along the fibre and the wave is lost.

The *aperture* is the diameter of the largest beam of light that can enter an optical fibre.

Parameters of an Optical Fibre

The parameters of an optical fibre are:

(a) The *specific refractive index* Δ. This is the specific refractive index given by equation (7.5):

$$\Delta = (\eta_1 - \eta_2)/\eta_1 \tag{7.5}$$

Typically Δ is about 0.01.

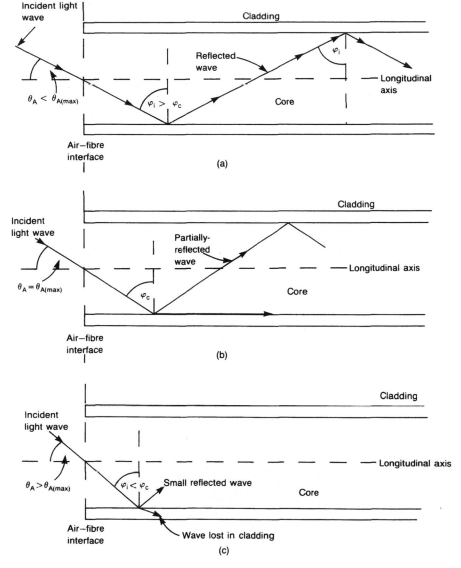

Fig. 7.6 Effect of acceptance angle on propagation in an optical fibre

(b) the *numerical index, NA,* which is equal to the sine of the maximum acceptance angle $\theta_{A(max)}$. It defines the acceptance cone shown in Fig. 7.5. The numerical index generally has a value somewhere in the range 0.2 to 0.4, giving a maximum acceptance angle of between 22° and 48°.

$$NA = \sin \theta_{A(max)} = \sqrt{(\eta_1^2 - \eta_2^2)} = \sqrt{[\eta_1^2 - \eta_1^2(1 - \Delta)^2]}$$
$$= \sqrt{(2\Delta\eta_1^2)} = \eta_1\sqrt{(2\Delta)} \qquad (7.6)$$

since $\eta_1^2\Delta^2$ is negligibly small.

(*c*) The *normalized frequency, f*, which specifies the number of different modes that may be propagated over a fibre; it is given by equation (7.7):

$$f = (2\pi a)NA/\lambda \qquad (7.7)$$

where *a* is the radius of the core.

Only one mode is able to propagate when *f* is less than 2.4; three modes can propagate if *f* is greater than 2.4 but less than 3.8, and so on for other values.

Example 7.2

An optical fibre has a core with a refractive index of 1.51 and a cladding whose refractive index is 1.48. Calculate (*a*) the numerical aperture, (*b*) the maximum acceptance angle and (*c*) the critical angle of incidence.

Solution

(*a*) $NA = \sqrt{(1.51^2 - 1.48^2)} = 0.3$ (*Ans.*)

(*b*) $\theta_{A(max)} = \sin^{-1} 0.3 = 17.5°$ (*Ans.*)

(*c*) $\varphi_c = \cos^{-1}(0.3/1.51) = 78.5°$ (*Ans.*)

This is an example of a *stepped-index* optical fibre. A stepped-index optical fibre is one that has a definite, sudden change in the refractive indices of the core and the cladding. The glass core will typically have a refractive index of about 1.5, while a typical refractive index for the cladding is about 1.8.

Dispersion

When a number of light waves enter the system and are incident upon the upper boundary with differing angles of incidence, but all greater than φ_c, a number of *modes* are able to propagate. Multimode propagation is shown by Fig. 7.7. There are a number of different paths over which the input light waves are able to propagate along the optical fibre. These paths are of varying lengths and therefore the light waves will take different times to pass over a given distance. This effect is known as *transit time dispersion* and it places a restriction on the maximum possible bit rate that can be transmitted over a stepped-index fibre.

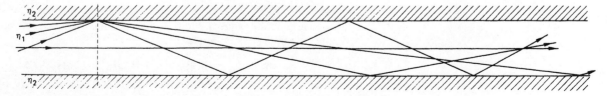

Fig. 7.7 Multimode propagation in a stepped-index fibre

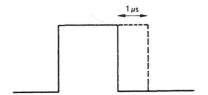

Fig. 7.8 Effect of transit time dispersion on a rectangular pulse

Suppose, for example, that a light wave following the shortest, direct, path through the fibre takes 5 μs to travel a certain distance, and that a light wave travelling over the longest path takes 6 μs to reach the same point. The effect on a rectangular pulse transmitted over the system is shown by Fig. 7.8. The trailing edge of the pulse is delayed in time by 1 μs. This effect will, of course, limit the maximum bit rate that is possible, since the leading edge of a following pulse must not arrive before the extended edge of the pulse shown has ended.

Dispersion in an optical fibre can be minimized in one of two ways.

(a) A *graded-index fibre* can be used. In this kind of fibre the refractive index of the core is highest at its centre and decreases gradually towards the edges. This variation ensures that the difference between the refractive indices of the outer part of the core and the cladding is small, and that the change can take place smoothly instead of abruptly as before. Light waves nearing the core—cladding interface will then be gradually refracted, rather than reflected, from the boundary as shown by Fig. 7.9. A wave entering the core at a large angle from the horizontal will penetrate a long way from the horizontal before it is refracted sufficiently to change its direction of travel. A wave entering the core at a shallower angle does not penetrate as far from the horizontal before it is refracted sufficiently to change its direction of travel.

Once a wave has been refracted back to the horizontal, it will enter the next section of the core at a shallower angle than before and it will not, therefore, travel as far from the horizontal before reversing its direction of travel. The effect of this is to reduce the differences between the lengths of the paths followed by the various light waves. Also, the waves that travel farthest from the axis are travelling in a lower refractive index medium and so they travel at a higher velocity than waves travelling near the axis. These effects ensure that all the light waves travelling through the core have almost the same transit time.

(b) A *single-mode fibre* can be used. If the diameter of the core is reduced to be of the same order of magnitude as the wavelength of the incident light wave, then only one mode will be able to propagate in the fibre. This is shown by Fig. 7.10.

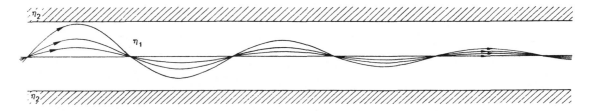

Fig. 7.9 Multimode propagation in a graded-index fibre

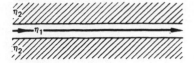

Fig. 7.10 Single-mode propagation in a stepped-index fibre

Fibre Optic Cable

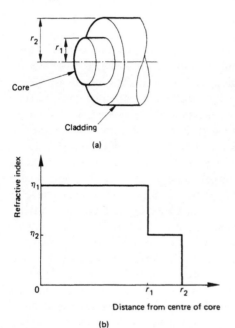

(a)

(b)

Fig. 7.11 (a) Stepped-index multimode optical fibre, (b) refractive index profile

No matter which of the three possible modes of propagation is used, the dimensions of the cladding must be at least several wavelengths at the frequency of the propagating light. Otherwise some light energy will be able to escape from the optical fibre, and extra losses will be caused by any light scattering and/or absorbing objects that are in the vicinity of the fibre. The available bandwidth is limited by (a) transit time dispersion and (b) material dispersion.

An optical fibre cable consists of a glass core that is completely surrounded by a glass cladding. The core performs the function of transmitting the light wave(s), while the purpose of the cladding is to minimize surface losses and to guide the light waves. The glass used for both the core and the cladding must be of very high purity since any impurities that are present will cause some scattering of light to occur. Two types of glass are commonly employed: silica-based glass (silica with some added oxide) and multi-component glass (e.g. sodium borosilicate).

There are three basic kinds of optical fibre cable.

(a) *Stepped-index Multimode*. The basic construction of a stepped-index multimode optical fibre is shown by Fig. 7.11(a) and its refractive index profile is shown by Fig. 7.11(b). It is clear that an abrupt change in the refractive index of the fibre occurs at the core–cladding boundary. The core diameter, $2r_1$, is often some $50-60$ μm but in some cases may be as large as 200 μm. The diameter $2r_2$ of the cladding is standardized, whenever possible, at 125 μm. Many tens of modes can simultaneously propagate.

(b) *Single-mode*. Figure 7.12(a) shows the basic construction of a single-mode optical fibre and Fig. 7.12(b) shows its refractive index profile. Once again the change in the refractive indices of the core and the cladding is an abrupt one but now the dimensions of the core are much smaller. The diameter of the core should be of the same order of magnitude as the wavelength of the light

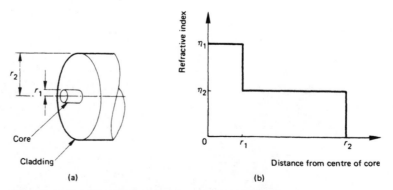

(a)   (b)

Fig. 7.12 (a) Single-mode optical fibre, (b) refractive index profile

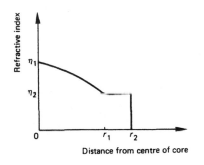

Fig. 7.13 Refractive index profile of graded-index multimode optical fibre

to be propagated and it is therefore in the range $1-10\ \mu\text{m}$. The cladding diameter is usually the standardized value of $125\ \mu\text{m}$.

(c) *Graded-index Multimode*. The basic construction of a graded-index multimode optical fibre is the same as that of the stepped-index multimode fibre shown in Fig. 7.11(a). The refractive index profile is not the same, however, and this is given by Fig. 7.13. The core has its greatest value of refractive index at its centre and this decreases parabolically towards the core–cladding interface. The core diameter is in the range $50-60\ \mu\text{m}$ and the cladding diameter is $125\ \mu\text{m}$.

The relative merits of the three kinds of optical fibre are as follows.

The stepped-index multimode fibre produces large transit time dispersion and in consequence it can only be employed for bandwidth-distance products of up to about 50 MHz km. The use of this kind of optical fibre is therefore restricted to low-speed applications such as data signals and various industrial control systems.

Single-mode optical fibre can provide very large bandwidth-distance products, of up to about 100 GHz km. Early difficulties in manufacture and jointing, and in the manufacture of coupling devices led to the development and use of graded-index optical fibre. Now that these difficulties have been overcome only single-mode fibre is installed in the core telephone network.

Graded-index multimode optical fibre has the ability to give bandwidth-distance products of 1 GHz km or more and, because of its larger dimensions, it is easier to manufacture and install.

Glass is a brittle material and an optical fibre cable must be protected against breakage both during its installation and over its anticipated lifetime. The necessary strength is provided by a steel wire situated in the middle of the cable and protection is given by a plastic sheath. The constructional detail of one type of cable is shown by Fig. 7.14. The steel wire at the centre of the cable has a bedding layer around it and then the six optical fibres are laid helically around the layer. The gaps between the fibres are filled up with a filling compound and then an aluminium-tape water-barrier is wrapped around to form

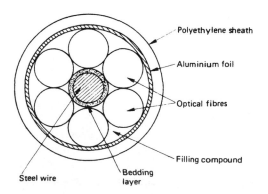

Fig. 7.14 Optical fibre cable

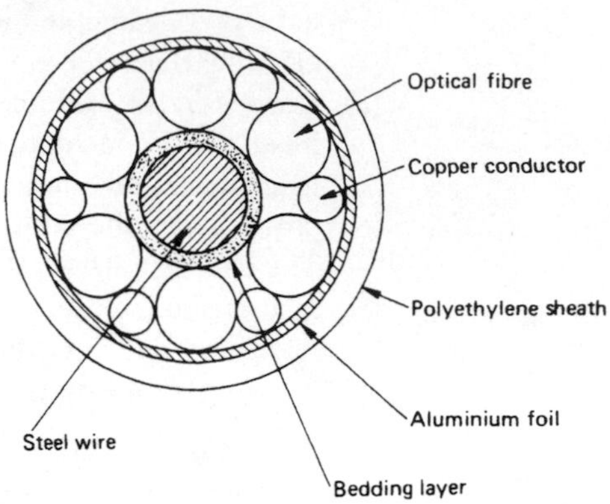

Fig. 7.15 Optical fibre cable with added copper conductors

the cable. Lastly, a polyethylene sheath is placed around the aluminium tape to complete the cable.

In some cases copper conductors are also provided within the cable; they may be used for power-carrying purposes and for speaker circuits. An example of such a cable is shown in Fig. 7.15.

Characteristics of Optical Fibre Cable

Launching Efficiency

The light power output of the light source has to feed into the optical fibre with the maximum possible efficiency. To this end, *couplers* have been designed to maximize the *launching efficiency*. The launching efficiency is the ratio

$$\eta = \frac{\text{power accepted by fibre}}{\text{power emitted by light source}} \times 100\% \tag{7.8}$$

Coupling losses of up to about 2 dB at each end of a fibre, i.e. 4 dB in total, are typical.

Attenuation

The light energy fed into an optical fibre is attenuated as it travels towards the far end of the fibre. The losses in an optical fibre are the sum of the contributions of a number of sources. The sources of loss are: (*a*) absorption, caused by metal and/or water impurities in the glass; (*b*) scattering in the core because of inhomogeneities in the refractive index − this is known as Rayleigh scattering; (*c*) scattering at the core−cladding boundary; and (*d*) losses due to radiation at each bend in the fibre.

The attenuation coefficient of an optical fibre refers only to losses in the fibre itself, i.e. neglecting coupling and bending losses.

The variation with frequency of the attenuation coefficient of an optical fibre depends upon the specific glass used for the core and the cladding. However, certain features are common to all types of fibre. The loss is high at about 1.7×10^{14} Hz and decreases rapidly with increase in frequency to reach a minimum at about 1.9×10^{14} Hz. Further increase in the frequency results in a gradual increase in the attenuation coefficient up to about 5×10^{14} Hz. However, some sharp peaks in the attenuation coefficient exist at approximately 2.16×10^{14} Hz and 2.42×10^{14} Hz.

Because of the very high frequencies involved it is customary to work in terms of wavelength rather than frequency. Figure 7.16 shows typical attenuation—wavelength curves for both single-mode and graded-index multimode optical fibres. The peaks at $1.24\,\mu$m and $1.39\,\mu$m occur because of excess absorption loss at these wavelengths. Low-loss windows occur in the characteristic of Fig. 7.16 at wavelengths of $1.3\,\mu$m and $1.5\,\mu$m and these are utilized for long-distance telecommunication systems; $1.3\,\mu$m for land systems and $1.5\,\mu$m for submarine systems.

Example 7.3

A laser diode launches 2 mW power into a single-mode optical fibre. Calculate the received power if the cable is 20 km in length and the wavelength used is 1.3 nm.

Solution

From Fig. 7.16 the attenuation of the cable is $0.4 \times 20 = 8$ dB. Hence,

$$P_{\text{out}} = 2\,\text{mW}/10^{0.8} = 2/6.31 = 0.317\,\text{mW} \qquad (Ans.)$$

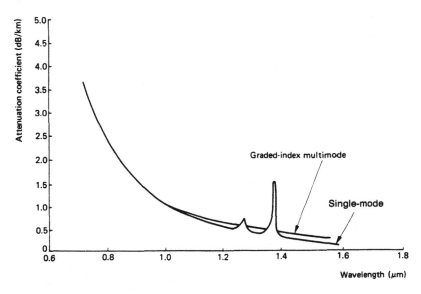

Fig. 7.16 Attenuation—wavelength characteristics of single-mode and graded-index multimode optical fibres

Velocity

The velocity of propagation of a light wave in an optical fibre is equal to the velocity of light divided by the refractive index of the core material. Thus,

$$\text{velocity} = (3 \times 10^8)/\eta_1 \qquad (7.9)$$

If, typically, $\eta_1 = 1.5$, then the velocity of propagation is 2×10^8 m/s.

Bandwidth

The bandwidth of an optical fibre is the number of megahertz that may be transmitted with an attenuation that is no greater than 3 dB over a 1 km length of the fibre. Bandwidth is quoted in MHz/km.

Optical Sources

The light source used in an optical fibre system must, of course, be able to deliver light energy at the appropriate wavelength. It must also be capable of being modulated at the very high bit rates used, with a low driving power and a relatively high output power. There are two kinds of light source which are able to satisfy these requirements; these are the *light emitting diode* and the *laser diode*.

The light emitting diode (LED) is generally used for short-distance, narrow-bandwidth systems, while the laser diode finds application in long-haul wide-bandwidth systems. The laser diode can develop about 100 times more output power than the LED can provide but it is both more expensive and less reliable at the present time.

The Light Emitting Diode

A semiconductor diode, fabricated using materials such as gallium arsenide (GaAs), gallium aluminium arsenide (GaAlAs), gallium indium arsenide phosphide (GaInAsP) or gallium indium arsenide (GaInAs), will emit light energy whenever it is forward biased and conducting a current. The emitted light is in the infrared part of the electromagnetic spectrum. The specific wavelength at which a light emitting diode (LED) radiates energy can be selected by the suitable choice of the semiconductor material. The wavelengths of the light radiated by four different materials are given in Table 7.1.

Table 7.1 LED wavelengths

Material	GaA1As	GaAs	GaInAs	GaInAsP
Wavelength of emitted light (μm)	0.8–0.9	0.9–1.0	1.0–1.1	1.25–1.35

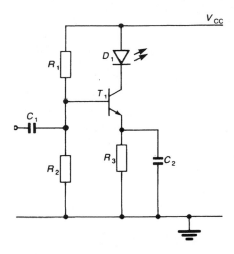

Fig. 7.17 LED drive circuit

There are two main types of LED available, one of which radiates the light energy from the surface of the device while the other radiates from the edge of the semiconductor structure. The LED may be used as the light source for telecommunication systems and produces non-coherent light output with a power of some 0.05 mW to 1 mW. The light is not of one wavelength but exhibits a *linewidth* or *spectral spread* that might be as much as 40 nm. Because of its non-coherent light output, the LED will emit light waves into an optical fibre at a variety of angles and it is therefore suitable for use with graded-index multimode fibres. There is a linear relationship between drive current and optical power output.

A typical LED drive circuit is shown in Fig. 7.17.

The Laser Diode

The construction of a laser diode differs from that of a LED mainly in that it incorporates a cavity which is resonant at the required frequency of the emitted light. When a voltage is applied across a laser diode it radiates *coherent* light energy. The spread in wavelengths is small, perhaps only 1–2 nm. Commonly, the semiconductor material employed is gallium aluminium arsenide and this gives a peak emission at a wavelength of 0.82 μm with more than 100 MHz bandwidth. The power output is in the range of 1 mW to 10 mW. Some commercially available laser diodes incorporate optical negative feedback to ensure a stable output over a wide range of ambient temperatures.

Laser diodes are usually ON–OFF modulated and are used for digital systems operating over single-mode optical fibre. Figure 7.18 shows a laser diode drive circuit. A PIN photodiode is used to monitor the light output and to provide negative feedback to keep the output constant to within about ±0.5 dB

The laser diode offers the following advantages over the LED: (*a*) small dimensions, (*b*) high efficiency, (*c*) it is easy to modulate, (*d*) directional emission, (*e*) output power ranges from a few milliwatts

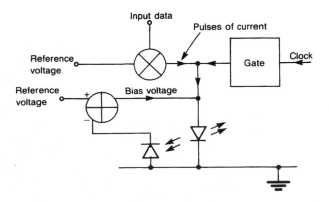

Fig. 7.18 Laser diode drive circuit

to about 10 W, and (*f*) a very short risetime (typically about 1 ns). On the other hand, the laser diode is more expensive, less reliable, and its performance is temperature sensitive. The laser diode finds its main applications in single-mode fibre systems operating mainly at a wavelength of 1.3 μm.

Optical Detectors

The function of an optical detector, or photodetector, is to convert input light energy into the corresponding electrical signal. The detector should have its maximum detection efficiency at the operating wavelength of the system and it should operate linearly at the modulated bit rate. Also, of course, it should be of physically small dimensions, and be both cheap and reliable. All of these requirements can be satisfied by the semiconductor photodiode.

When a reverse-biased p—n junction is illuminated by light energy, a current will flow across the junction whose magnitude is very nearly directly proportional to the illumination intensity. The current flowing is very nearly independent of the actual bias voltage so long as the p—n junction is kept reverse biased. If the diode current is passed through a load resistance, a detected output voltage can be obtained.

Two kinds of photodiode are employed: the silicon *PIN photodiode* and the silicon *avalanche photodiode*. The PIN photodiode is so named because it has a region of intrinsic semiconductor sandwiched in between n-type and p-type regions. This device is suitable for the shorter wavelengths but, for wavelengths longer than about 1.1 μm, its sensitivity is no longer good enough and then either a GaInAs or a GaInAsP device would be employed instead. The silicon avalanche photodiode is essentially a PIN device also but it is fabricated in such a way as to accentuate the *avalanche effect*. The avalanche photodiode can provide a large current gain as well as detection but it has a tendency to be rather noisy.

Optical Passive Devices

Optical passive devices include directional couplers, switches, isolators, and attenuators.

An *optical directional coupler* is used to split and combine an optical signal. The basic principle of operation of the device is illustrated by Fig. 7.19. It has three ports, each of which is coupled to a different optical fibre. Signals entering the directional coupler at port 1 can be directed to either port 2 or port 3, or, conversely, the signals entering at ports 2 and 3 can be combined to leave at port 1. An optical switch can be used to switch a light wave from one path to another. The switch includes a prism whose position can be toggled between two positions by a d.c. control pulse applied to the device.

An *optical isolator* will tranmit light with very little loss in one direction, but in the other direction will block the signal to give at least 25 dB attenuation. The concept of an isolator is illustrated by Fig. 7.20.

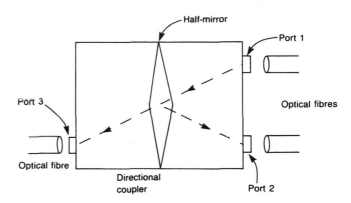

Fig. 7.19 Optical directional coupler

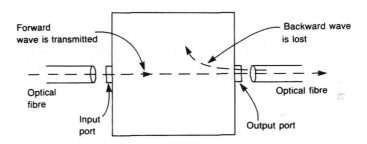

Fig. 7.20 Optical isolator

An *optical attenuator* introduces attenuation into a light path by causing some of the input light to be reflected away from the output port. The basic principle of operation of an attenuator is shown by Fig. 7.21; the input light is incident on a glass substrate on which has been deposited a metallic film. The glass substrate is mounted at an angle to the optical path so that the reflected light does not return to the input port. Both fixed-value, continuously variable and stepped-value attenuators can be obtained with losses that may go as high as 60 dB.

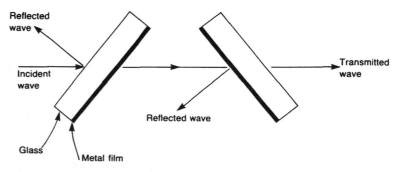

Fig. 7.21 Optical attenuator

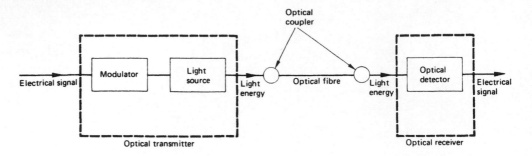

Fig. 7.22 Basic block diagram of an optical fibre system

Optical Fibre Telecommunication Systems

The basic block diagram of an optical fibre telecommunication system is shown by Fig. 7.22. The electrical signal to be transmitted over the system is fed into the optical transmitter. Here it is applied to the modulator to modulate the light source. The modulated light energy is coupled to the optical fibre and is then transmitted to the far end using one of the modes of propagation described earlier, single-mode for modern systems. At the optical receiver, incoming light energy is detected in order to convert the signal back to its original electrical format.

The characteristics of the light source make it best suited to some form of digital modulation in which the source will be switched ON and OFF. The potential bandwidth offered by an optical fibre is very wide and usually the input electrical signal consists of the output from a PCM system.

The HDB3 coded output signal of the PCM system is not suitable for transmission over an optical fibre and must be converted into a more suitable format, such as CMI, before transmission. Figure 7.23 shows the block diagram of a typical system. The HDB3 signal is scrambled to produce a purely random pulse train and thus avoid any possibility of repetitive patterns occurring in the transmitted digital signal. The encoded signal is then applied to the modulator and this modulates the laser diode to shift the PCM signal to the allocated optical wavelength. The light signal produced by the diode is then coupled into the optical fibre cable. At the receiver the incoming light signal is detected by the optical detector and converted back into electrical form. The electrical signal is then amplified and equalized before it is applied to a pulse regenerator. The regenerated output signal is then decoded to restore the signal to its original HDB3 format.

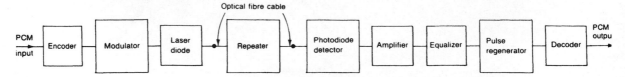

Fig. 7.23 System with optical pulse regenerator

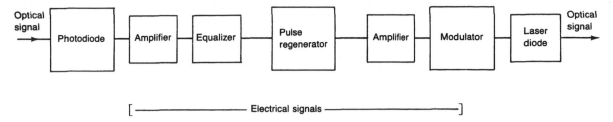

Fig. 7.24 Optical pulse regenerator

If the system length is in excess of about 40 km it will be necessary to regenerate the transmitted light pulses to restore their wave-shape and this is the function of a *repeater*. The block diagram of a repeater is shown in Fig. 7.24.

Wavelength-division Multiplex

The use of *wavelength-division multiplex* increases the information carrying capacity of a single optical fibre by allowing the simultaneous transmission of two, or more, wavelengths. Unlike frequency-division multiplex or time-division multiplex, wavelength-division multiplex need not involve the use of active devices. At each end of a system the combination and separation of the multiplexed light signals is carried out by optical filters. The basic principle of wavelength-division multiplex is shown by Fig. 7.25(a). A glass plate that is coated with a substance that makes it optically transparent at wavelength λ_1 and optically reflective at wavelength λ_2 is mounted at 45° to the optical signal path. Light waves entering the multiplexer at ports 1 and 2 are then directed to the output port 3. Signals entering port 1 are at wavelength λ_1 and these are able to pass straight through the glass plate and onto output port 3. Light waves that enter the device

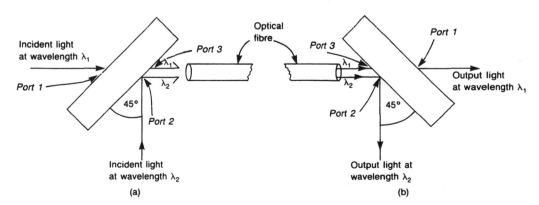

Fig. 7.25 Wavelength-division multiplex

at port 2 are reflected from the glass plate and also directed to output port 3. At the other end of the optical system the reverse process takes place to separate the two signals and direct them to the appropriate ports; this is shown by Fig. 7.25(*b*).

Exercises

7.1 A ray of light travels through a medium of refractive index 1.68. Calculate the minimum angle of refraction that will give total internal refraction.

7.2 Draw, and briefly explain the operation of, the block diagram of a pulse regenerator for use in an optical fibre system.

7.3 State the wavelengths that are employed for optical fibre communications and say which kind of fibre is used in each case. List two light sources and two light detectors for these wavelengths.

7.4 List the advantages of optical fibre systems over coaxial cable systems.

7.5 The repeaters in a conventional coaxial cable system need to be spaced at approximately 2 km intervals along a route. The repeaters in an optical fibre system are spaced at about 40 km intervals. List, and explain, the advantages arising from this greater repeater spacing.

7.6 Describe how light energy can be propagated along an optical fibre by means of reflection and refraction. If, for a particular fibre, the cladding has a refractive index 1% less than the refractive index of the core, calculate the critical angle of incidence.

7.7 Explain, with the aid of a sketch, what is meant by multimode propagation in a stepped-index fibre. How can this effect lead to transit time dispersion? Determine the maximum bit rate possible if 1 μs wide pulses are used and if the transit time of the longest path is 1 μs longer than that of the direct path.

7.8 Draw the block diagram of an optical fibre transmission system. Describe the function of each of the blocks drawn. State what kind of modulation is employed.

7.9 List the sources of loss in an optical fibre system. Briefly explain each of them. Draw a typical attenuation—wavelength characteristic and say why particular wavelengths only are so far used for optical fibre systems.

7.10 The operating wavelength of an optical fibre system is 0.85 μm. Determine the dimensions of the cable cores shown in Figs 7.11(*a*) and 7.12(*a*) in terms of wavelengths.

7.11 An optical fibre cable has a bandwidth-distance product of 2 GHz km. Determine (*a*) the maximum frequency that can be used over a distance of 20 km, (*b*) the maximum cable length possible for transmission at 500 MHz.

7.12 Explain the terms stepped-index, graded-index and single-mode when applied to an optical fibre. Draw, and explain, the refractive index profiles of each fibre.

8 Noise

The output of a communication system, be it line or radio, will always contain some unwanted voltages or currents in addition to the desired signal. The unwanted output signals are known as *noise* and they may have one or more of a number of different causes. These signals are classed as being either external noise or internal noise. External noise is induced into the equipment from an external source and internal noise is generated within the equipment itself.

For the signal received at the end of a system to be of use, the signal power must be greater than the noise power by an amount depending upon the nature of the signal. The ratio of the wanted signal power to the unwanted noise power is known as the *signal-to-noise ratio*. The signal level must never be allowed to fall below the value that gives the required minimum signal-to-noise ratio, because any gain introduced thereafter will increase the level of both the noise and the signal by the same amount and so it will not improve the signal-to-noise ratio.

Noise having a constant energy per unit bandwidth over a particular frequency band (uniform noise power density) is said to be *white noise*. Similarly, if the noise level decreases with increase in frequency at the rate of 3 dB/octave it is *pink*, while a noise power–frequency slope of 6 dB/octave gives *red* noise. (Note that the terms pink and red noise are not accepted by all workers in the field.) *Impulse noise* occurs in relatively high amplitude pulses that have a short duration compared with the time interval between the pulses.

The sources of noise are many; some noise is generated by various mechanisms within the equipment while further noise may by picked up by aerials and lines.

Sources of Noise

Thermal Agitation Noise

If the temperature of a conductor is increased from absolute zero (-273 °C), the atoms of the conductor begin to vibrate and some electrons are able to break away from their parent atoms and wander freely within the conductor. The amplitude of the atomic vibrations increases with increase in the temperature of the conductor and so the number of free electrons also increases with temperature. The free electrons then wander in a random manner within the conductor, but at any particular instant more electrons are travelling in some directions than in others. The movement of an electron constitutes the flow of a minute current, and therefore the net current represented by the moving electrons fluctuates continuously in both its magnitude and its direction. Over a period that is long compared with the average time for which an electron travels in a particular direction, the total current is zero. The continuous flow of minute currents develops a random voltage across the conductor, and this unwanted voltage is known as *thermal agitation* noise or as *resistance* noise.

The r.m.s. noise voltage produced by thermal agitation in a conductor is given by

$$V_n = \sqrt{(4kTBR)} \tag{8.1}$$

where k = Boltzmann's constant = 1.38×10^{-23} J/K
T = temperature of conductor K*
B = bandwidth (Hz) over which noise is measured, or of circuit at whose output the noise appears, whichever is the smaller. For most practical purposes, B can be taken as the 3 dB bandwidth of the circuit
R = resistance of circuit in ohms

Equation (8.1) may also be extended to find the noise voltage produced in an impedance; R is then the resistive component of that impedance.

It is the bandwidth and not the frequency of operation that is important with regard to thermal agitation noise. Thus a wide-band amplifier is noisier than a narrow-band amplifier, whatever their operating frequencies may be.

Example 8.1

Calculate the noise voltage produced in a 78 kΩ resistance in a 2 MHz bandwidth if the temperature is 20 °C.

Solution
From equation (8.1),

$$V_n = \sqrt{(4 \times 1.38 \times 10^{-23} \times 293 \times 2 \times 10^6 \times 78 \times 10^3)}$$
$$= 50.2 \, \mu V \quad (Ans.)$$

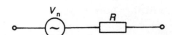

Fig. 8.1 Noise voltage generated in a resistance

The thermal noise e.m.f. may be regarded as acting in series with the resistance R producing it as shown by Fig. 8.1.

Resistors in Series

The total noise e.m.f. generated by two, or more, resistors connected in series is determined by adding their mean square values and then taking the square root of the sum. Figure 8.2 shows three resistors

* K = °C + 273

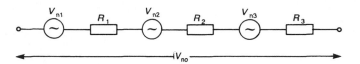

Fig. 8.2 Noise voltages in resistors in series

connected in series and their equivalent noise generators. The total mean square noise voltage V_{no}^2 is

$$V_{no}^2 = V_{n1}^2 + V_{n2}^2 + V_{n3}^2 = 4kTB(R_1 + R_2 + R_3)$$

and

$$V_{no} = \sqrt{[4kTB(R_1 + R_2 + R_3)]} \qquad (8.2)$$

This means that the total noise voltage is the same as would be generated in a resistance whose value was equal to the sum of the three resistors.

Example 8.2

Resistors of 22 kΩ, 56 kΩ and 30 kΩ are connected in series. Calculate the total noise generated at 20 °C in a 1 MHz bandwidth.

Solution
$V_{n1} = \sqrt{(4 \times 1.38 \times 10^{-23} \times 293 \times 10^6 \times 22 \times 10^3)} = 18.86\,\mu\text{V}$
$V_{n2} = 18.86\sqrt{(56/22)} = 30.09\,\mu\text{V}$
$V_{n3} = 18.86\sqrt{(30/22)} = 22.02\,\mu\text{V}$
$V_{no} = \sqrt{(18.86^2 + 30.09^2 + 22.02^2)} = 41.8\,\mu\text{V} \qquad (Ans.)$

Alternatively,

Total resistance $= 108$ kΩ
$V_{no} = \sqrt{(4 \times 1.38 \times 10^{-23} \times 293 \times 10^6 \times 108 \times 10^3)} = 41.8\,\mu\text{V}$
$(Ans.)$

Similarly, the total noise voltage generated in resistors connected in parallel can be found by determining their total resistance and then using this value in equation (8.2).

Available Noise Power

Maximum power transfer from a resistive source to a load occurs when the load resistance is equal to the resistance of the source. Consider a resistance R connected across another resistance of the same value that may be considered to be noiseless (Fig. 8.3). The noise power delivered to the load resistance is

$$P_a = (V_n/2)^2/R = (4kTBR)/4R = kTB \text{ watts} \qquad (8.3)$$

Thus the maximum or *available noise power* that can be delivered by a resistance is independent of the value of that resistance but is proportional to *both* temperature *and* bandwidth. It is often convenient to note that, if the temperature is 290 K (17 °C), the available noise power is 4×10^{-15} W/MHz. Thermal agitation noise is *white*.

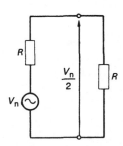

Fig. 8.3 Available noise power

Example 8.3

Calculate the available noise power from a resistance at a temperature of 17 °C over a 2 MHz bandwidth.

$$P_a = 2 \times 4 \times 10^{-15} = 8 \times 10^{-15}\,\text{W} \quad (Ans.)$$

Noise in Semiconductors

The output current of a bipolar transistor, a FET or an integrated circuit consists of a d.c. determined by the d.c. operating conditions, plus a superimposed a.c. current that is determined by the input signal current or voltage. Random fluctuations in these currents always exist, however, which may be considered to be the result of the superimposition of a noise current on the d.c. and signal currents.

Noise in Bipolar Transistors

THERMAL AGITATION NOISE
Thermal agitation noise is generated in all three regions of a bipolar transistor, but particularly in the base.

SHOT NOISE
Shot noise in a transistor is caused by random fluctuations in the numbers of holes and electrons crossing each p−n junction. Since there are two p−n junctions in a transistor there are two sources of shot noise.

Shot noise is given by

$$I_n = \sqrt{(2eI_{DC}B)} \tag{8.4}$$

where I_{DC} is the d.c. collector current, B is the bandwidth, and e is the electronic charge $1.602 \times 10^{-19}\,\text{C}$.

PARTITION NOISE
The input current to a bipolar transistor flows through the emitter to the base−emitter junction. After crossing the junction it divides between the collector and base terminals ($I_E = I_B + I_C$). This current division is also subject to random fluctuations and so it is another source of noise.

$1/f$ NOISE
Fluctuations in the conductivity of the semiconductor material produce a noise source which is inversely proportional to frequency. This noise, also known as kicker noise, is usually negligible above about 1 kHz.

Noise in FETs

Noise in a FET originates from three sources: shot noise generated by leakage currents in the gate−source p−n junction, thermal noise generated in the channel resistance, and $1/f$ noise caused by the random generation and recombination of charge carriers. The FET is inherently a lower noise device than is a bipolar transistor, although if a bipolar transistor is operated with a d.c. collector current of only a few microamperes its noise performance may be superior.

The reasons for the generally superior noise performance of a FET are: (*a*) that its structure contains only one p−n junction as opposed to two in a bipolar transistor (this means that shot noise is less); and (*b*) that the current flowing into the source can only flow out of the drain and this means that the FET is not subject to partition noise.

Noise in Line Systems

The noise output of a line communication system consists of the sum of the noise generated in the transmission line itself and the noise produced within the terminal stations, and the pulse regenerators along the route.

The noise arising in the transmission line is the sum of thermal agitation noise in the line resistance, noise due to faulty joints, interference picked up from nearby power lines or electric railways, and crosstalk from other pairs in the same cable. Thermal agitation noise has already been considered and noise caused by a faulty joint needs no discussion.

If a transmission line runs more or less parallel to a power line or an electric railway, it may have unwanted power-frequency voltages induced in it via inductive and/or capacitive couplings between the lines. Underground cables often have a metallic sheath that acts as a screen to reduce the magnitude of the unwanted voltages. Coaxial pairs are generally operated with their outer conductor earthed and are quite efficiently self-screened. This type of interference is minimized by keeping telecommunication cables spaced as far away from power lines as possible.

Crosstalk is a voltage appearing in one pair in a cable when a signal is applied to another pair. Any multi-pair cable will experience crosstalk between all its pairs to a greater or lesser extent. Crosstalk in a cable is caused by electrical couplings between the conductors; these couplings may be capacitive, magnetic or via insulation resistances. The construction of a multi-pair cable is designed to minimize crosstalk, and, when necessary, balancing the couplings between pairs at the end of each section of line can give a further reduction. Crosstalk is not a problem in optical fibre cables.

In both repeater stations and telephone exchanges, noise voltages are introduced by thermal agitation in the equipment, by faulty connections, by inadequate smoothing of the power supplies, by IC, bipolar transistor and FET noise, and by crosstalk. Crosstalk occurs in the internal wiring because of capacitive and inductive couplings between wires. In electronic equipment crosstalk is the result of electric couplings between inadequately screened components and couplings via the common power supplies. Crosstalk via the power supplies can occur because the current taken by each circuit flows in the common internal resistance of the power supply and develops a voltage across it. To minimize crosstalk in power supplies they are designed to have very low internal resistance.

Impulse noise, generated by a wide variety of sources, such as electric motors or switches, may be picked up by wiring. Data circuits are more sensitive to noise and interference than speech circuits. For instance, short breaks in the transmission path of a millisecond or so would be unnoticed in a speech circuit, but could cause considerable error in a data circuit. Such breaks in the transmission path would probably be the result of poorly made joints in cable terminations and at distribution frames.

In addition to the various sources of noise previously mentioned, data circuits are also adversely affected by impulse noise, such as dial impulses transmitted in one cable pair appearing, as a result of crosstalk, in another cable pair. Such interference is troublesome if the bit duration is short in comparison with the time duration of the dial pulses. This source of noise will vary in intensity with the amount of telephone traffic and so it will vary with the time of day and the day of the week. A noise impulse of 1 ms duration could affect several bits in a data waveform to cause an increase in the bit error rate. Thus:

1200 bits/s; bit duration $\simeq$ 0.83 ms   1 to 2 bits corrupted
2400 bits/s; bit duration $\simeq$ 0.42 ms   2 to 3 bits corrupted
4800 bits/s; bit duration $\simeq$ 0.21 ms   5 bits corrupted
9600 bits/s; bit duration $\simeq$ 0.1 ms   10 bits corrupted.

Quantization Noise

Another form of noise which occurs in PCM multi-channel telephony systems is known as *quantization noise*. It arises because of the way that sampled voltages are rounded off to the nearest of a number of admissible voltage, or quantum, levels. Quantization noise is discussed in Chapter 12.

Noise in Radio Systems

Radio receivers are subject to the same sources of internal noise as electronic circuits and line equipment. The noise picked up by an aerial originates from a number of different sources which may be divided into natural sources and man-made sources.

Natural Sources of Noise

Noise is picked up by an aerial from the sky and also from the earth itself. At frequencies in the MF and HF bands atmospheric noise, or *static*, is always present because it is generated by thunderstorms. Several thunderstorms are always in progress at any one time somewhere on the Earth's surface and since such noise is able to propagate for very long distances some static is present at any one point at all times. The audible effect at the output of a radio receiver is very similar to that of thermal agitation noise. Atmospheric noise is most significant at frequencies in the region of 15 kHz but it falls to a negligible level at frequencies above about 20 MHz. Noise is also picked up by aerials that point towards the sky, from both the sun and distant stars, from about 20 MHz to about 500 MHz. The effect of solar noise can always be reduced by making sure that the aerial does not point towards the sun.

Man-made Sources of Noise

Whenever an electrical current is switched on or off noise voltages are generated. These noise voltages are known as *radio frequency interference* (RFI) above 30 MHz, and as *electromagnetic interference* (EMI) between 30 kHz and 30 MHz. EMI may travel over the mains wiring for long distances and RFI may be conducted over the mains wiring and be radiated from it. Man-made noise can often be suppressed at its source and whenever possible this is done.

The sources of man-made noise are many and include brushes on electrical motors, neon lights, electrical switches and fluorescent lights.

Signal-to-Noise Ratio

The output of a communication system, whether line or radio, will always contain some unwanted voltages or currents in addition to the desired signal. The unwanted output signal is known as noise and may have one or more of a number of different causes. For the signal received at the end of a system to be of use, the signal power must be greater than the noise power by an amount depending upon the nature of the signal. The ratio of the wanted signal *power* to the unwanted noise *power* is known as the signal-to-noise ratio, i.e.

$$\text{Signal-to-noise ratio} = \frac{\text{wanted signal power}}{\text{unwanted noise power}} \qquad (8.5)$$

or

$$\text{Signal-to-noise ratio} = 10 \log_{10}\left(\frac{\text{wanted signal power}}{\text{unwanted noise power}}\right) \text{dB} \qquad (8.6)$$

The signal-to-noise ratio required of a particular system depends upon the potential use of the signal and is generally determined by means of subjective tests. For example, a line music circuit may require a signal-to-noise ratio of 60 dB in order that the transmitted music may not be noticeably degraded, but a telephone circuit only requires about 35–40 dB. The required signal-to-noise ratio determines the spacing of the relay stations in a microwave radio-relay link, and it is a factor in the minimum transmitter power necessary for a radio system. For economic reasons, therefore, it is necessary to maximize the signal-to-noise ratio of a system by reducing the magnitudes of any noise sources as far as possible.

Example 8.4

The signal voltage at the output of an amplifier is 1.2 V and the noise voltage that is unavoidably also present is 12 mV. Calculate the signal-to-noise ratio at the output of the amplifier.

Solution
Signal-to-noise ratio is the ratio of the wanted signal *power* to the unwanted noise *power*. Therefore

$$\text{Signal-to-noise ratio} = 1.2^2/(12 \times 10^{-3})^2 = 10\,000 \text{ or } 40 \text{ dB}$$
$$(Ans.)$$

Most naturally occurring noise sources produce a noise power which is directly proportional to bandwidth. This means that the signal-to-noise ratio at the output of an amplifier, a radio receiver, or some other kind of electronic, line or radio equipment is inversely proportional to the bandwidth of that equipment.

Example 8.5

The signal-to-noise ratio at the output of a radio-frequency amplifier is 1000. What would be the signal-to-noise ratio if the bandwidth of the amplifier were doubled?

Solution
If the bandwidth of the amplifier is doubled the noise power at the output terminals will also be doubled. The signal output power is unchanged and so the signal-to-noise ratio will be reduced by half to 500.

Example 8.6

An amplifier has a gain of 30 dB and generates a noise power, referred to its input terminals, of 3 μW. If the signal applied to the amplifier input is −10 dBm* with a signal-to-noise ratio of 20 dB, calculate the signal-to-noise ratio at the output of the amplifier.

Solution
Output signal level = −10 + 30 = +20 dBm
Input noise level P_N = −10 − 20 = −30 dBm
$30 = 10 \log_{10}[(1 \times 10^{-3})/P_N]$
$10^3 = 1000 = (1 \times 10^{-3})/P_N$, or

$$P_N = 1 \, \mu\text{W}$$

The total amplifier noise, referred to the input, is 3 μW, and hence the total input noise power is 4 μW. In dBm,

*dBm: decibels relative to 1 milliwatt.

$$x = 10 \ \log_{10}[(4 \times 10^{-6})/(1 \times 10^{-3})] = 10 \ \log_{10}(4 \times 10^{-3})$$
$$= -24 \ \text{dBm}$$

The output noise power is thus $-24 + 30 = +6$ dBm and the output signal-to-noise ratio is

$$20 - 6 = 14 \ \text{dB} \quad (Ans.)$$

The noise generated in the line resistance will degrade the signal-to-noise ratio by an amount equal to its attenuation. If, therefore, a signal having a signal-to-noise ratio of 80 dB is applied to a line having a loss of 20 dB, the signal-to-noise ratio at the output of the line will be reduced to 60 dB. This figure will *not* be improved by amplifying the signal since *both* signal *and* noise will be amplified to the same extent. Instead, because of the various sources of noise in an amplifier, the signal-to-noise ratio will be further degraded.

In *any* analogue transmission system the signal-to-noise ratio must always progressively worsen as the length of the system is increased. This is to be contrasted with a digital system such as pulse code modulation in which noise is *not* cumulative with distance.

Exercises

8.1   An amplifier has a rectangular bandwidth of 100 kHz and a voltage gain of 150. Calculate the noise voltage at its output due to a 78 kΩ resistor connected across its input terminals. Will this be the total noise voltage at the output terminals? Assume the temperature to be 20°C and $k = 1.38 \times 10^{-23}$ J/K.

8.2   List and briefly explain the sources of noise in bipolar and field effect transistors.

8.3   Calculate the thermal noise voltage generated by a 500 kΩ resistor if the bandwidth is 2 MHz and the temperature is 17°C. $k = 1.38 \times 10^{-23}$ J/K.

8.4   Calculate the available noise power from a resistor in W/MHz if the temperature is 18°C. $k = 1.38 \times 10^{-23}$ J/K.

8.5   Explain why data circuits are more susceptible to noise than speech circuits. Give three sources of noise that may affect a data circuit but, unless severe, will not affect a speech circuit.

8.6   Calculate the signal-to-noise ratio at a point where the signal voltage is 10 V and the noise voltage is 100 mV.

8.7   The noise power level at a point where the signal-to-noise ratio is 37 dB is 100 μW. Calculate the signal power.

8.8   Calculate the noise power developed by a 56 kΩ resistor in a bandwidth of 100 kHz at a temperature of 17°C.

8.9   The noise level at the input to an amplifier is -40 dBm. If the input signal-to-noise ratio is 30 dB determine the amplifier gain necessary to give an output signal of $+17$ dBm.

8.10   The input noise to an amplifier is given by kTB watts. Calculate the input signal level needed to give an input signal-to-noise ratio of 45 dB. $T = 290$ K; $B = 2$ MHz.

9 Digital Signals and their Transmission over Lines

In the past telephone networks have been primarily designed and planned for the transmission of analogue speech signals. This appears to be the simplest and most straightforward method of working since the speech signal need only be band-limited and amplified at suitable intervals along the line. Short lines, such as local lines connecting the user to his local telephone exchange, or trunks interconnecting nearby telephone exchanges, will require no processing at all.

A number of advantages can be gained, however, if the speech signal is converted into a representative binary-coded digital signal before transmission over the network. The process employed to achieve this is known as *pulse code modulation* (PCM) and is considered in Chapter 12. The digital signal transmitted to line may be of the form shown

Fig. 9.1 Unipolar digital signal

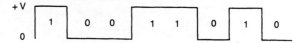

in Fig. 9.1. Each pulse of positive voltage represents binary 1 and times when no pulse is transmitted indicate binary 0. This is an example of a *unipolar* digital signal. Often, binary 0 is represented by a negative voltage pulse, as shown by Fig. 9.2, to give a *bipolar* digital signal.

Fig. 9.2 Bipolar digital signal

The d.c. data signals produced by a computer or a data terminal are of similar form and generally use the International Alphabet 5; this is also known as the American Code for Information Interchange (ASCII) (see Appendix A). In this code each character is represented by seven bits followed by an eighth *parity* bit, each of which may be either a 1 or a 0. An example of the ASCII code is given in Fig. 9.3; it shows the characters B and R using *non-synchronous synchronization*. At the end of each character the data signal always returns to the binary 1 voltage level for a period of time at least equal to the

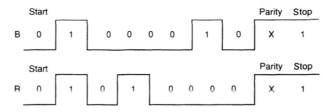

Fig. 9.3 ASCII waveforms
(X ≡ 0 or 1)

time duration of a single bit. The signal will remain in the stop state until such time as the next character is to be transmitted; this is indicated by the line voltage changing from the binary 1 to the binary 0 voltage level for a one-bit period before the seven bits representing the character begin.

The change in state from 1 to 0 of the start bit of each character indicates the start of a sampling period for the distant receiver; it indicates to the receiver that it must commence sampling the incoming bits one time interval later. It does this by starting the clock in the receiver that synchronizes the sampling instants. Thereafter, the receiver continues to sample the received bits during each of the following six time intervals (ideally in the middle of each). At the end of the last bit in a character the stop bit is received; this resets the receiver clock so that it is able to repeat the process when the next start bit arrives.

Although non-synchronous start-stop synchronization works and it is relatively simple to implement, it is inefficient since three in every ten bits convey no real information. A more efficient, although more complex, system is the synchronous transmission of data. With a synchronous system start and stop bits are not employed. Instead, the receiver clock runs continuously and is kept in synchronism with the transmitter clock by synchronization pulses transmitted along with the data.

Digital signals cannot be transmitted over a telephone network, for other than very short distances, without some kind of processing. The resistance and capacitance of the line combine to distort the received waveform, and any line transformers, amplifiers and other equipments will remove the d.c. component (the average value) from the digital waveform. As a result, for other than very short distances, either pulse regenerators or modems must be employed.

The intention of most, if not all, telephone administrations is to convert their networks from analogue to wholly digital operation using pulse regenerators. The use of digital transmission instead of analogue transmission gives the following advantages.

(*a*) Better signal-to-noise ratio; this is because noise and interference are not cumulative with distance in a digital system as they are in an analogue system.
(*b*) Improved signalling.
(*c*) Multiplexing using TDM is simpler than multiplexing using FDM.

(*d*) Digital switching is easier to implement.

(*e*) Different kinds of digital signal, e.g. data, telegraph, telephone speech and television, can be treated as identical signals during transmission and switching. This is the concept behind the new system of operation of telephone networks, known as Integrated Services Digital Network (ISDN), that is just being introduced into networks throughout the world. Because of the economics involved the change from analogue to digital networks will take some time to implement.

The main disadvantage of digital operation is the wider bandwidths needed, but the main reason for its employment being fairly recent is that the widespread introduction of many LSI integrated circuits has made complex circuit functions economic to carry out. The ISDN is a single digital network that is able to transmit and switch a number of different communication services. An all-digital line network, based upon PCM telephony systems, is fully integrated with computer-controlled switching equipment in digital telephone exchanges. An ISDN will be able to handle data, slow-scan television, teletext and telemetry signals as well as telephony speech signals with equal facility.

Unipolar and Bipolar Operation

The digital signal that is transmitted to line can be either a unipolar or a bipolar signal. Unipolar operation means that only one polarity of voltage is used, and data is signalled by the application of, and the removal of, the voltage applied to the sending-end terminals of the line. With bipolar operation two voltages are employed, one being positive and the other negative with respect to earth.

Data circuits can be operated in any one of the simplex, the half-duplex, or the full-duplex modes. *Simplex* means that transmission of data is possible in only one direction over the link. A *half-duplex* circuit can be operated to transmit in either direction but only in one direction at a time. Lastly, a *full-duplex* circuit is capable of simultaneously transmitting data in both directions.

The principle of a unipolar simplex circuit is illustrated by the circuit shown by Fig. 9.4(*a*). The electronic switch connects either +6 V or earth potential to the line. When +6 V is applied a voltage pulse travels along the line to the receiver connected across the far-end terminals. At the receiver the pulse is interpreted as indicating binary 1. When earth potential is applied to the line there is a period of time during which no voltage pulse is received by the receiver and this is taken as indicating binary 0. Loop-disconnect dialling, used by older telephones, also employs unipolar d.c. signals.

At zero and low frequencies the inductance and the leakance of the line both have negligible effect upon the signal and only the line's resistance and capacitance will affect the waveform of the received current. Figure 9.5 is an approximate representation of a low-

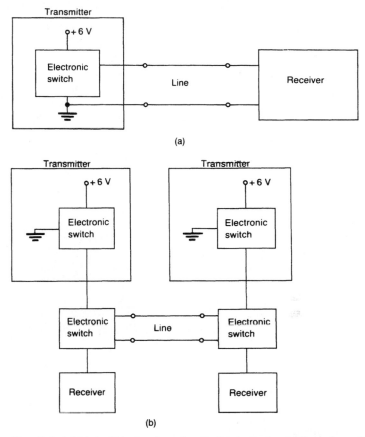

(a)

(b)

Fig. 9.4 (*a*) Unipolar simplex circuit, (*b*) unipolar half-duplex circuit

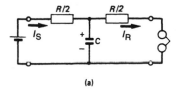

(a)

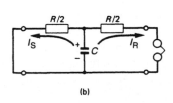

(b)

Fig. 9.5 Low-frequency equivalent circuit of a transmission line showing the currents flowing for unipolar operation

frequency line which assumes that the total shunt capacitance is concentrated at the centre of the line. When the d.c. voltage is connected to the sending end of the line, most of the current which flows into the line is used to charge the line capacitance. The current arriving at the far end of the line increases at a relatively slow rate and it is unable to reach its final *steady-state* value until the line capacitance has been fully charged. Once the line capacitance is fully charged, the sending-end current falls until it reaches the value needed to maintain the received current at its steady-state value. The time required for the received current to reach its steady-state value is proportional to the *total* capacitance C and resistance R of the line. Similarly, when the d.c. voltage is removed from the sending end of the line and the input terminal is earthed, the line capacitance will start to discharge at a rate determined by the time constant CRl^2, where l is the length of the line in kilometres. The discharge current is in the *opposite* direction to the original sending-end current, hence the sending-end current reverses its direction (see Fig. 9.5(*b*)). Some of the discharge current flows out of the receive end of the line in the *same* direction as the earlier current and so prolongs the received

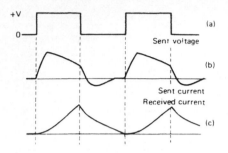

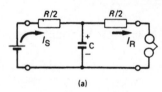

Fig. 9.6 Unipolar waveforms

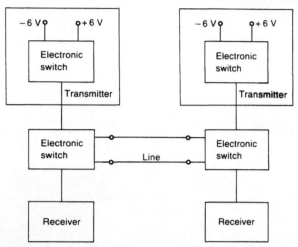

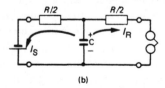

Fig. 9.7 Bipolar half-duplex circuit

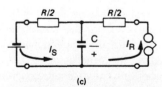

Fig. 9.8 Low-frequency equivalent circuit of a transmission line showing the currents flowing for double-current operation

current. Because of this the received current does not fall as rapidly as the sent current, neither does it reverse its direction. Figure 9.6 gives the waveforms of (a) the sent voltage, (b) the sent current, (c) the received current. The *risetime* of the output current is the time taken for the current to increase from 10% to 90% of its final value.

Figure 9.4(b) shows the arrangement of a unipolar half-duplex circuit; the operation is very similar to that just described except that transmission is possible in either direction but not both simultaneously.

Figure 9.7 shows the basic circuit of a half-duplex bipolar circuit. The operation of the transmitter reverses the polarity of the voltage applied to the sending end of the line. Suppose a positive voltage, with respect to earth, is first applied to the line (Fig. 9.8(a)). When the line capacitance has been fully charged, the received current will reach its steady-state value after a time determined, as before, by the time constant CRl^2.

When the polarity of the sending-end voltage is first reversed to become negative (Fig. 9.8(b)), the line capacitance will start to discharge through both the transmitter and the receiver. The polarity of the reversed applied voltage is such that it acts in the same direction as the voltage developed across the line capacitance. As a result the capacitance of the line discharges much more rapidly than in the unipolar case and hence the received current falls more rapidly. Once the line capacitance has completely discharged, it is then charged in the opposite direction; most of the sending-end current is used to charge the line capacitance and thus the received current increases, in the opposite direction to before, at a relatively slow rate, and it does not reach its steady-state value until the line capacitance has been fully charged. This is shown by Fig. 9.8(c).

Waveforms of current and voltage for a bipolar circuit are given in Fig. 9.9. Clearly, the risetime and the falltime of the received current waveform are smaller than in the unipolar case and the amplitude of the received current is greater.

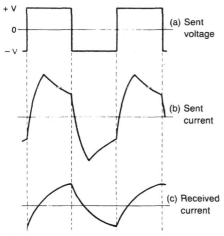

Fig. 9.9 Double-current waveforms

The end of the circuit that is transmitting at any given instant connects either $+6\,V$ or $-6\,V$ to the line. Binary 1 is indicated by $-6\,V$ and binary 0 is indicated by $+6\,V$. Although the voltages have been shown as $\pm 6\,V$ in the figure, the relevant Electronic Industries Association (EIA) standard allows voltages from $\pm 3\,V$ to $\pm 25\,V$ to be used.

Relative Merits of Unipolar and Bipolar Working

Unipolar operation of a d.c. data circuit requires only one-polarity voltage supply, whereas bipolar operation requires the provision of both positive and negative power supplies.

If the bit rate of the data waveform is too high, the time taken for the received current to reach its steady-state value may exceed the duration of the pulse. Excessive pulse distortion will then occur, making it difficult for the receiver to reliably interpret received pulses as either 0 or 1.

If the current received at the end of a cable pair is not allowed sufficient time to reach its maximum value (steady-state) before the signal ends, the waveform of the transmitted signal will not be reproduced. The resultant waveform will not contain all the frequencies predicted from a knowledge of the fundamental frequency of the data waveform. If the time taken for the received current to build up to its steady-state value is less than the time duration of a bit, the received current waveform will only be affected by the varying loss of the line at different frequencies. If the risetime of the received current is greater than the bit length, distortion will occur.

Some examples are given in Fig. 9.10. When the time taken for the received current to reach its steady-state value is equal to the bit duration, the received current waveform is as shown by Fig. 9.10(b). Clearly, the current waveform is not rectangular.

If the received current is able to reach its steady-state value before the trailing edge of the next voltage pulse occurs, its waveshape is approximately rectangular (see Fig. 9.10(c)). Conversely, should the pulse duration be much shorter than the time needed for the received current to attain its steady state, the current will never reach its steady value and considerable waveform distortion will occur (Fig. 9.10(d)).

Double-current operation of a data link increases the rate at which the received current rises towards its final value and so allows a higher bit rate to be employed. Double-current working also results in a larger-amplitude received current than does the single-current system for the same applied d.c. voltage. This gives more reliable operation.

A further disadvantage of the single-current method of operation is that any momentary break in the transmission path will not be detected as such but will be recorded as binary 0 and so produce an error in the received data.

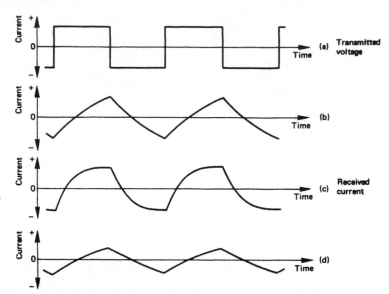

Fig. 9.10 (a) Showing the effect of the risetime time of the received current on the received current waveform; (b) risetime time equal to bit duration; (c) risetime time less than bit duration; (d) build-up time greater than bit duration

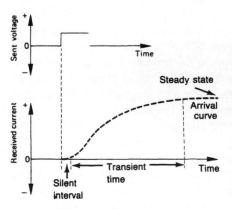

Fig. 9.11 Arrival curve for a single pulse

The Arrival Curve

The *arrival curve* is a graph of the received current at the end of a line plotted against time for each positive or negative pulse taken separately, ignoring for simplicity any propagation delays (as was done in the previous figures). The arrival curve for a single pulse is shown by Fig. 9.11. The arrival curve can be used to construct the waveform of the received current for any input data waveform. This can be used to determine the time which is required for the received current to attain the value at which the receiver operates. The silent interval is equal to the time it takes the pulse to propagate over the line. It is the delay that occurs between a voltage pulse being applied to one end of the line and the current starting to increase at the other end of the line.

The part of the arrival curve labelled as the transient time is the time taken for the received current to reach its steady-state value. Each input voltage pulse is assumed to produce a current at the receiving end of the line corresponding to the arrival curve. The arrival curve for each input pulse is separately drawn and the waveform of the actual received current is then deduced by adding algebraically the individual arrival curves.

The sketches given in Fig. 9.12 apply this principle to, firstly, two pulses of differing time durations and, secondly, a square waveform, both signals being centred on 0 V.

The Frequencies Contained in a Digital Waveform

The data fed into and out of a computer uses both positive and negative voltage pulses to represent the binary numbers 0 and 1, as shown by

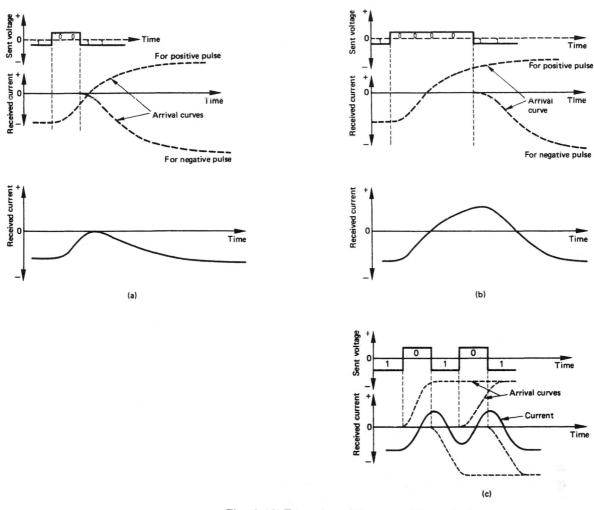

Fig. 9.12 Examples of the use of the arrival curve

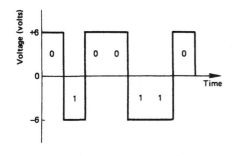

Fig. 9.13 Digital waveform

Fig. 9.13. Each signal element or *bit* has the same time duration in seconds as all the other bits and the number of bits transmitted per second is known as the *bit rate*. If, for example, the bit duration in Fig. 9.13 is 0.2083 ms, the bit rate would be $1/(0.2083 \times 10^{-3})$ or 4800 bits/s.

Figure 9.14 shows a sinusoidal voltage $v = V \sin \omega t$, of peak value V and frequency $\omega/2\pi$, and another, smaller, voltage at three times the frequency, i.e. $3\omega/2\pi$. The first voltage is the fundamental frequency and the other voltage is its third harmonic. At time $t = 0$, the two voltages are in phase with one another and the waveform produced by the summation of their instantaneous values is shown as a broken line; clearly the resultant waveform is tending towards a rectangular shape.

If the fifth harmonic, also with zero phase angle at time $t = 0$, is

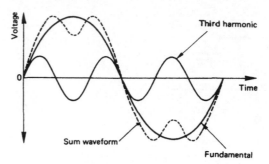

Fig. 9.14 Waveform produced by a fundamental and its third harmonic

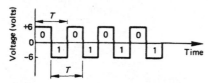

Fig. 9.15 Square digital waveform

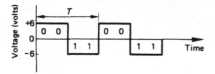

Fig. 9.16 A square digital waveform of lower bit rate

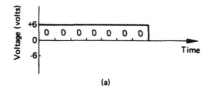

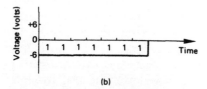

Fig. 9.17 Digital waveforms with zero fundamental frequency

added to the fundamental and the third harmonic, the resultant waveform is more nearly rectangular. Adding the seventh, ninth, etc. odd harmonics produces an even better approximation to the rectangular waveshape. If all the odd harmonics up to a very high order (theoretically infinity) are included, a pulse train of square waveform is obtained (see Fig. 9.15).

The number of pulses occurring per second is known as the *pulse repetition frequency* (PRF) and it is equal to the fundamental frequency contained in the pulse waveform. The *periodic time T* of the waveform is the reciprocal of the PRF and is the time interval between the leading edges of consecutive pulses.

A digital waveform will only be square when it consists of alternate 0s and 1s as shown in Fig. 9.15. In the periodic time T, one 0 followed by one 1 occur, i.e. two bits. The fundamental frequency of this waveform is equal to $1/T$ and it is equal to one-half the number of bits per second. If the digital waveform consists of alternate pairs of 0s and 1s as shown by Fig. 9.16, four bits occur in the periodic time T of the waveform and so the fundamental frequency of the waveform is now equal to one-quarter of the bit rate.

When the digital waveform is made up of a number of consecutive 0s or 1s as shown by Fig. 9.17, the digital voltage is constant and so the frequency of the waveform is zero hertz.

The data transmitted over a link will include all the various combinations of 0s and 1s, according to the information content, and so the fundamental frequency produced will vary from a minimum of zero hertz to a maximum equal to one-half the bit rate. The more rapidly the data waveform changes, or in other words the higher the bit rate, the higher will be the frequencies of its harmonic components. The minimum bandwidth that must be provided is equal to one-half the bit rate. If only this minimum bandwidth is provided, the digital waveform would lose its rectangular shape to become sinusoidal since none of its harmonics would be transmitted.

Example 9.1

Determine the maximum fundamental frequency of a 2400 bits/s data waveform. What other frequencies are present?

Solution

Maximum fundamental frequency = 2400/2 = 1200 Hz.

Other frequencies present are

(i) Third harmonic 3600 Hz
(ii) Fifth harmonic 6000 Hz, etc.

The Effect of Lines on Digital Signals

The attenuation of a telephone line increases with increase in frequency and so the various harmonics contained in a digital waveform will be attenuated to a greater extent than the fundamental frequency. The greater attenuation suffered by the harmonics, particularly the higher-order ones, means that the rectangular waveshape of a d.c. data signal will be lost. This effect is accentuated as the length of the line and hence its attenuation is increased, with the result that the pulses become more and more rounded as they travel along a line. This effect is shown by Fig. 9.18. The higher the bit rate, the higher the fundamental frequency of the digital waveform and the shorter will be the length of line needed to reach the point at which satisfactory reception is impossible. Reception is, of course, affected by noise picked up by the line. The equipment at the receiving end of the line will sample each incoming bit at its mid-point and, provided it is able to determine reliably at any instant in time whether a 1 or a 0 is present, reception will be satisfactory.

Fig. 9.18 Effect of line attenuation on a transmitted digital waveform

Because of the effect of line attenuation, direct transmission of data waveforms over unamplified telephone cables is only possible at bit rates of up to 150 bits/s, and then only for fairly short distances of up to about 4 km. At these bit rates, the effect of group delay—frequency distortion is negligibly small.

If the distance between them is *very* short, two computers can be directly connected together without interface equipment and be able to communicate with one another at a high bit rate. The arrangement is shown by Fig. 9.19.

The signals generated by digital equipment cannot be transmitted over a link set up via the PSTN and/or over an amplified circuit

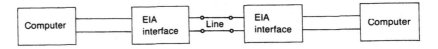

Fig. 9.19 Direct interconnection of two computers

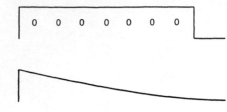

Fig. 9.20 Showing the effect of removing d.c. component: (a) transmitted, (b) received

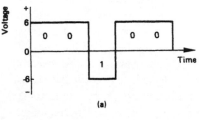

(a)

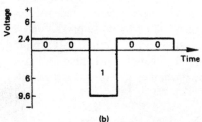

(b)

Fig. 9.21 Showing the effect of removing the d.c. component from a digital waveform: (a) with d.c. component, (b) without d.c. component

because the d.c. component of the signal will be lost and the data waveform altered. (The d.c. component of a data waveform is the average value.) Further, if the bit rate is low the fundamental frequency and, perhaps, some of the lower-order harmonics may also not be transmitted. There are four reasons why the PSTN cannot transmit the d.c. or very low frequency components of a data waveform.

(a) The telephone exchange switching equipment.
(b) Line matching transformers that may be used in the access network.
(c) Many links will be routed over one or more multi-channel telephony systems, or a radio link, and these neither provide a d.c. path nor pass frequencies below 300 Hz.

If the digital waveform consists of alternate 1s and 0s, its d.c. component will be zero (Fig. 9.15). At the other extreme when the digital waveform consists of a number of consecutive 1s, its d.c. component is −6 V; similarly, when consecutive 0s are sent, the d.c. value is +6 V. In either of these cases, the removal of the d.c. component would result in the effect shown in Fig. 9.20. Clearly, a loss of information is very likely.

Many other combinations of 1s and 0s are also transmitted in digital waveforms and, if their d.c. component is lost, the waveform will be altered. Suppose, for example, the digital waveform is that shown by Fig. 9.21(a); the d.c. component, or average value, of this waveform is 3 × 6/5 volts (since the 1 pulse cancels out one 0 pulse) or +3.6 V. If this d.c. component is removed from the waveform, the resulting waveform will vary from a positive voltage of +6 − 3.6 = +2.4 V to a negative voltage of −6 − 3.6 = −9.6 V, as shown in Fig. 9.21(b). The effects of line attenuation and noise will soon ensure that the receiving data terminal will be unable to reliably detect the binary 0 pulses.

There are two methods of overcoming these effects: either the digital signal can be converted to a voice-frequency signal, using digital modulation, that is transmitted in the normal commercial-quality speech bandwidth, or the pulses can be regenerated at regular intervals along the line.

The Use of Modems in Digital Systems

Figure 9.22 shows an analogue voice-frequency point-to-point data circuit. Two *modems* are used to convert the d.c. data signals to voice-

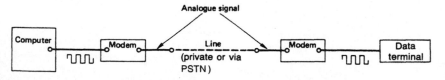

Fig. 9.22 Use of modems for data communication

frequency signals at the transmitter in one direction of transmission, and to convert the signals back to their original d.c. form at the receiver. The digital signal is used to modulate a suitable carrier frequency and the modulated (voice-frequency) signal is transmitted over the circuit to the distant end. At the distant end another modem demodulates or detects the incoming signal to recover the digital signal. The link may be a point-to-point circuit leased permanently from the telephone company, known as a *dedicated circuit*, or it may be a dialled connection set up via the PSTN.

The Use of Pulse Regenerators in Digital Systems

The principle of pulse regeneration is illustrated by Fig. 9.23. The pulses transmitted to the line by the digital terminal suffer distortion, because of the combined effects of line attenuation and group delay—frequency distortion and noise, but they are recreated to their original waveform by each line pulse regenerator. The data waveform is reconstituted without error provided the pulse waveform has not been degraded to such an extent that the pulse regenerator is unable to determine reliably at each instant in time the presence or absence of a pulse. Effectively, any noise and distortion accompanying the signal is removed by the pulse regenerator.

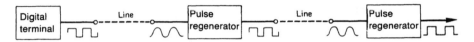

Fig. 9.23 Use of pulse regenerators

Synchronization

When two digital terminals are linked together and interchanging information, the terminal receiving information must be *synchronized*.

Synchronization is essential so that the receiver will, at all times, sample each incoming bit at the correct instant in time. Otherwise, the possibility exists that one, or more, bits may be lost with catastrophic effects on the accuracy of the received data. Suppose, for example, that the three denary numbers 13, 17 and 15 are transmitted using 5-bit binary code. The transmitted data is then 011011000101111 (printed in the order of transmission, left-hand side first). If, because of a lack of synchronization, the initial bit is missed, the received data would become 11011000101111 or 27, 2 and either 31 or 30 (depending on whether the next bit to appear is a 1 or a 0).

Synchronization in a data system is achieved either synchronously or non-synchronously.

Non-synchronous Data Transmission

In a non-synchronous data circuit the clock in the receiver is generated internally and it is kept in synchronism with the transmitter clock by the use of start and stop bits. When the circuit is idle the line voltage is held by the transmitter at the binary 1 level and the receiver clock is stopped. When the transmitter has a character to send it first switches the line voltage to the binary 0 voltage level for a one-bit time period. This is known as the *start bit*. The 8-bit character (seven data bits plus a parity bit) is then transmitted. The clock in the receiver is started by the leading edge of the start bit and it then generates the clock pulses that are used in the receiver to determine the instants at which the incoming data bits are sampled. The synchronization of the receiver should be such that each data bit is sampled at its mid-point. When the last (parity) bit has been transmitted, a *stop bit*, at the binary 1 voltage level, is transmitted. This stop bit stops the receiver clock until such time as the next start bit is received. This means that synchronization of the receiver clock to the transmitter clock is carried out on a character-by-character basis. There is no need for the receiver clock to be very stable because there is insufficient time for any clock inaccuracies to produce significant error. A crystal oscillator is often employed as the clock. Figure 9.24(*a*) shows a non-synchronous data signal. Each of the data bits may be either a 1 or a 0 depending on the actual character being transmitted. The parity bit is used to enable the receiver to detect any errors which may occur in the received data and it also may be either a 1 or a 0. Each character has a total length of 10 bits.

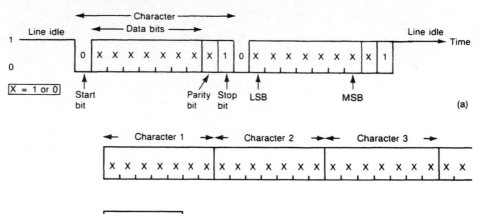

Fig. 9.24 (*a*) Non-synchronous and (*b*) synchronous data signals

Synchronous Data Transmission

With synchronous transmission blocks of data are continuously transmitted without either start or stop bits. The clock in the receiver runs continuously and it is synchronized to the transmitter clock by clocking information that is transmitted along with the data. Figure 9.24(*b*) shows a synchronous data bit stream; it is usually preceded by some synchronization bytes that the receiver is able to recognize. The timing of both the receiver and the transmitter is controlled by a clock in either the transmitting data terminal or its associated modem. The receiving modem and data terminal derive their timing information from the incoming data itself and/or synchronization bits inserted into the data stream. In one case, for example, the receiver compares the incoming bits with the receive clock, and if need be adjusts the clock to minimize any error. The synchronizing bits are far fewer than those required for a non-synchronous system and so the system is more efficient in its transfer of actual data. Synchronous systems work at bit rates of 1200 bits/s and upwards.

Synchronization is also needed in a digital telephone network and this is discussed in Chapter 12.

Signalling Systems

Clearly, there is a need for signalling systems in a telephone network to make it possible for a customer, or an operator, to establish a connection and, later, to clear the connection.

Older telephones originate a call using loop-disconnect signalling, in which the action of a telephone dial is to break, and make, the line loop to the local telephone exchange. The loop is broken a number of times that corresponds to the number dialled. Thus, if number 3 is dialled, three break and make pulses are generated. In the UK the dial break period is set at 67 ms and the make period at 33 ms; this means that 10 pulses (number 0) occupy one second. Loop-disconnect signalling is limited in its range by pulse distortion caused by the capacitance of the local line, since a dialled pulse train is an example of a unipolar waveform (see Fig. 9.6).

Modern telephones use a keypad to call another telephone. Each time a key is pressed a combination of two tones is sent to line. The tones used are given in Table 9.1

Table 9.1

Digit	Tones (Hz)		Digit	Tones (Hz)	
1	697	1209	7	852	1209
2	697	1336	8	852	1336
3	697	1477	9	852	1477
4	770	1209	*	941	1209
5	770	1336	0	941	1336
6	770	1477	#	941	1477

One advantage of digital transmission using PCM is that all the signalling requirements for 30 channels can be satisfied in one time slot. This is considered in Chapter 12.

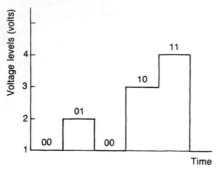

Fig. 9.25

When binary signals are transmitted over a channel at a bit rate of r bits/s the maximum fundamental frequency occurs for alternate 1s and 0s and is equal to $r/2$ Hz. The minimum bandwidth that must be provided is equal to the maximum fundamental frequency, i.e. $B = r/2$ Hz. With this bandwidth r bits/s is the maximum information transmission rate or the *channel capacity*.

If a four-level signal is transmitted in which each level represents a dibit, Fig. 9.25, the capacity is increased to $2r$ bits/s. Doubling the number of signalling levels to eight, so that each level represents a tribit, increases the capacity to $4r$ bits/s and so on.

For an n-level signal the channel capacity will be

$$C = r \log_2 n \text{ bits/s} \tag{9.1}$$

or

$$C = 2B \log_2 n \text{ bits/s} \tag{9.2}$$

If there was zero noise in a channel it would be possible continually to increase the capacity by employing more and more signalling levels (at the expense, of course, of more complex terminal equipment). In practice, some noise is *always* present in every channel. The effect of this noise on the bit error rate worsens as the distance between adjacent signal levels is reduced.

If the signal power and the noise power at the receiving end of a channel are S and N, respectively, the output signal-to-noise ratio is S/N. The r.m.s. output voltage will be $\sqrt{(S + N)}$ volts, while the r.m.s. noise output voltage will be $\sqrt{N}$ volts. To keep the probability of error to an acceptably low figure, the distance between adjacent signalling levels must be at least $\sqrt{N}$. This means that the maximum number n of possible signalling levels is $n = \sqrt{(S + N)}/\sqrt{N} = \sqrt{(1 + S/N)}$. Substituting into Equation (9.2)

$$C = 2B \log_2\sqrt{(1 + S/N)} \text{ bits/s} \tag{9.3}$$

or

$$C = B \log_2(1 + S/N) \text{ bits/s} \tag{9.4}$$

Example 9.2

Calculate the capacity of a channel of bandwidth 3000 Hz and output signal-to-noise ratio 30 dB.

Solution
30 dB is a power ratio of 1000:1. Therefore

$$C = 3000 \log_2(1 + 1000)$$

$$= 3000(\log_{10}10^3)/(\log_{10}2) = 29\,897 \text{ bits/s} \quad (Ans.)$$

This figure is the theoretical maximum capacity of the channel but to attain it would require the use of $n = \sqrt{(1 + 1000)} \simeq 32$ levels. This is far in excess of any number that is practically possible. If binary signals were to be employed, for example, the maximum bit rate would be only $2B = 6000$ bits/s. Clearly, the capacity of a channel can be increased by increasing its bandwidth and/or its output signal-to-noise ratio. Conversely, for a given value of channel capacity a trade-off between bandwidth and signal-to-noise ratio can always be made.

Example 9.3

Calculate the capacity of a channel of bandwidth 1200 Hz and signal-to-noise ratio 20 dB. If the bandwidth is reduced to 1000 Hz what is then the minimum allowable signal-to-noise ratio?

Solution

$$C = 1200 \log_2(1 + 100) = 7990 \text{ bits/s}$$

When the bandwidth is reduced to 1000 Hz,

$$7990/1000 = \log_{10}(1 + S/N)/\log_{10}2$$

$$\log_{10}(1 + S/N) = 2.4$$

and

$$\text{Signal-to-noise ratio} = 251 = 24 \text{ dB} \qquad (Ans.)$$

Exercises

9.1   The arrival curve for a particular line has a silent interval of 10 μs, a transit time of 30 μs, and a steady current of 4 mA for 6 ms. Sketch the curve.

9.2   The Morse code signal dot, dash, dot is sent over a line using (a) unipolar and (b) bipolar operation. If the time duration of a dash is three times that of a dot sketch the sent current waveform for each case.

9.3   (a) Explain, with the aid of suitable sketches, how the waveform of the current at the end of a digital data line link can be constructed using the arrival curve. (b) Use the arrival curve method to determine the waveform of the received current at the end of a line when the data waveform shown in Fig. 9.26 is applied to the sending end of the line.

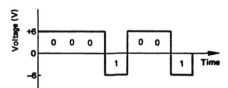

Fig. 9.26

9.4   (a) Draw graphs of current and voltage against time at the sending and receiving ends of a digital data line when (i) unipolar and (ii) bipolar signals are used.
(b) Use your graphs to explain why bipolar working gives greater reliability and a faster speed of signalling.

9.5 List the relative merits of unipolar and bipolar working of a short digital data link.

9.6 Draw the digital data waveform which represents the binary number 011001.

9.7 A data signal is transmitted at 4800 bits/s. A character consists of four 1 bits followed by four 0 bits. Calculate the fundamental frequency of the waveform.

9.8 A 4800 bits/s signal is transmitted by a computer. What are the (a) minimum and (b) maximum fundamental frequencies of the waveform? What other frequencies may also be present?

9.9 Draw a sine wave of frequency 300 Hz and peak value 6 V. Then draw the third and fifth harmonics of this wave with peak values of 2 V and 1 V respectively. Sum the three waves to obtain the resultant waveform.

9.10 The duration of each bit on a line is 1.042×10^{-4} s. Calculate the bit rate.

9.11 Explain why long-distance signalling employs voice-frequency signals instead of d.c. even though the latter are generated by the telephone dial.

9.12 Explain how group delay–frequency distortion has an effect on (a) analogue and (b) digital signals.

9.13 List, and briefly explain, four items in a telephone network that will not transmit the d.c. component of a data waveform.

9.14 (a) The digital data waveform shown in Fig. 9.27 has its d.c. component removed. Sketch the resultant waveform.
(b) Why would the d.c. component of a digital data waveform be removed if the signal were transmitted over the PSTN?

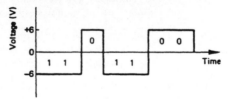

Fig. 9.27

9.15 (a) Explain, with the aid of waveform diagrams, the effects of attenuation and group delay–frequency distortion on a digital data waveform.
(b) Show, with the aid of block diagrams, how the effects of (a) can be overcome using (i) modems and amplifiers, (ii) regenerators.

10 Attenuators, Equalizers and Filters

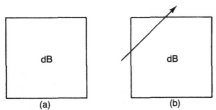

Fig. 10.1 Symbols for (a) a fixed attenuator, (b) a variable attenuator

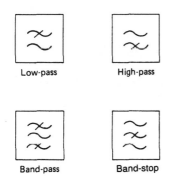

Fig. 10.2 Symbols for types of filter network

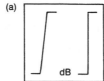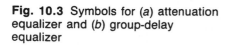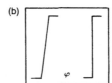

Fig. 10.3 Symbols for (a) attenuation equalizer and (b) group-delay equalizer

In both line and radio systems the need often arises for the level of a signal to be reduced, or attenuated, for a group of frequencies contained within a wider band to be transmitted while all other frequencies are suppressed, or for the attenuation−frequency or the group delay−frequency characteristic of a line to be altered.

An *attenuator* is a circuit that produces a given amount of attenuation to all the frequencies contained in a signal. The symbols for fixed and variable attenuators are given by Fig. 10.1(a) and (b), respectively.

A *filter* is a circuit that has the ability to discriminate between signals at different frequencies because it has an attenuation that varies with frequency in a particular manner. If a signal containing components at a number of different frequencies is applied to the input of a filter, only some of those components will appear at its output terminals, the remainder having been greatly attenuated, and so effectively suppressed. Four basic types of filter are available for use in telecommunication systems: the low-pass, the high-pass, the band-pass and the band-stop. Figure 10.2 shows the circuit symbols for each of these filters. Most filters can be designed using one of the following different techniques: inductor−capacitor filter, crystal filter and active filter.

An *equalizer* is a circuit that is fitted to the end of a line and has the function of making either the overall attenuation−frequency characteristic, or the group delay−frequency characteristic constant over a given frequency band. The symbol for an attenuation equalizer is shown in Fig. 10.3(a) and the symbol for a group-delay equalizer is given by Fig. 10.3(b).

Attenuators

An attenuator is a resistive network whose function is to provide a specified amount of loss or attenuation when it is connected between specified source and load resistances. The input and output terminals of the attenuator are then *matched* to the source and to the load.

Generally, one of three network configurations is employed; these are, respectively, the T, the π and the L networks. The three networks are shown in Fig. 10.4. Besides providing a specified loss the networks will also match the load to the source.

If the source and load resistances are of equal value a symmetrical network must be used. This means that the L network cannot be employed, and for both the T and π networks resistor R_1 must be of the same value as resistor R_3. The common value of the source resistance and the load resistance is then the same as the input and output resistances of the network and this value is known as the *characteristic resistance* R_0 of the attenuator.

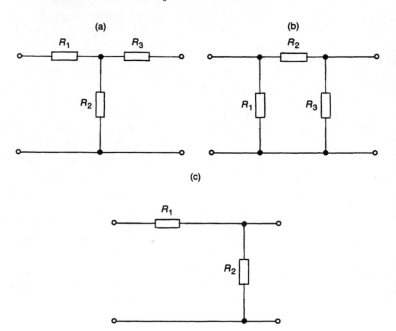

Fig. 10.4 (a) T attenuator, (b) π attenuator and (c) L attenuator

The necessary resistance values for a symmetrical attenuator to provide a specified attenuation are easily calculated using either equation (10.1) or equation (10.2).

T network: $R_1 = R_0(N-1)/(N+1)$, $R_2 = 2R_0N/(N^2-1)$

$$(10.1)$$

π network: $R_1 = R_0(N^2-1)/2N$, $R_2 = R_0(N+1)/(N-1)$

$$(10.2)$$

In these equations N represents the ratio

[input current (voltage)]/[output current (voltage)]

Example 10.1

Design a T attenuator to have a loss of 6 dB and a characteristic resistance of 600 Ω. Calculate the actual loss if the nearest preferred values are employed.

Solution

$20 \log_{10} N = 6$ dB, hence $N = 2$.

From Equation 10.1,

$$R_1 = 600(2 - 1)/(2 + 1) \quad = 200 \, \Omega \quad \quad (Ans.) \quad \text{(A preferred value)}$$

$$R_2 = (2 \times 600 \times 2)/(4 - 1) = 800 \, \Omega \quad \quad (Ans.)$$

The nearest preferred value is 820 Ω.

The circuit is shown by Fig. 10.5 from which:

$$\begin{aligned}
R_{\text{in}} &= 200 + (820 \times 800)/(820 + 800) = 605 \, \Omega \\
I_{\text{in}} &= (E/1205) \\
I_{\text{out}} &= [(E/1205) \times 820]/(820 + 200 + 600) \\
&= E/2381 \\
\text{Loss} &= 20 \log_{10}[(E/1205)/(2381/E)] = 5.92 \, \text{dB} \quad \quad (Ans.)
\end{aligned}$$

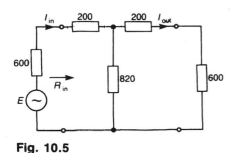

Fig. 10.5

If two or more attenuators are connected in cascade their overall attenuation, in decibels, is the arithmetic sum of their individual attenuations, in decibels, *provided* the attenuators are matched, i.e. they have the same values of characteristic resistance.

If an attenuator is not matched to either its source or its load or both, the loss obtained will be different from its attenuation by an amount that is not easy to determine. Therefore it is usual to attempt to match attenuators to both their source and their load resistances.

Equalizers

It was seen in Chapter 6 that both the attenuation and the group delay of a transmission line vary with frequency. If the line characteristics are not equalized the signal waveform at the end of the line will be distorted.

An *attenuation equalizer* is a circuit that is fitted at the end of a line as shown by Fig. 10.6. The component values of the equalizer are chosen so that the equalizer has a loss—frequency characteristic

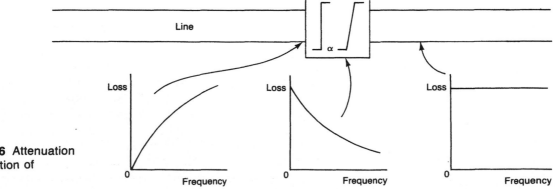

Fig. 10.6 Attenuation equalization of a line

which is the inverse of that of the line. The overall loss of the line *plus* the equalizer is the sum of their individual losses at each frequency, and so the overall loss—frequency characteristic is more or less constant. The process is illustrated by the loss—frequency curves given in Fig. 10.6.

The overall loss of the line is increased by the equalization process, particularly at the lower frequencies, but this is of little account since the equalized signal can easily be amplified. Two equalizer circuits are shown in Figs 10.7(a) and (b).

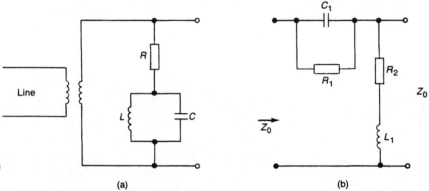

Fig. 10.7 (a) AF equalizer, (b) constant-impedance equalizer

(a)

(b)

At low frequencies the inductor L in Fig. 10.7(a) has little effect and the basic loss of the circuit is provided by the resistor R. With increase in frequency the impedance of the LC parallel circuit increases and so the shunt loss introduced by the equalizer falls. For better results more complex equalizers are employed; these are examples of *constant-impedance* equalizers, see Fig. 10.7(b). Much more complex circuitry is used in modem equalizers.

The function of a *group-delay equalizer* is similar to that of the attenuation equalizer. The equalizer is fitted at the end of the line and is set to have a group delay—frequency characteristic which is the inverse of the line's (see Fig. 10.8). Since the group-delay equalizer is required to have a frequency-dependent phase shift and, ideally, no loss, it is constructed using inductors and capacitors only. A wide variety of group-delay equalizer circuits are possible and Fig. 10.9 gives one possible example.

Inductor—Capacitor Filters

The transmission of an unwanted frequency through a network can be prevented either by connecting a high impedance (at that frequency) in series, and/or by connecting a low impedance in shunt, with the signal path. The high series impedance will oppose the flow of currents, at the unwanted frequencies, through the network, and the shunt impedance will bypass unwanted currents to earth. The necessary high and low values of impedance can be obtained by the use of inductors and capacitors of appropriate value. This is because an

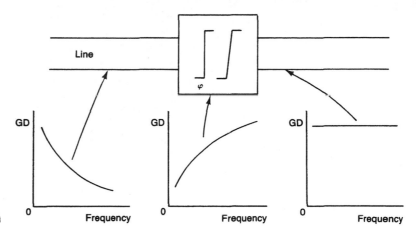

Fig. 10.8 Group-delay equalization of a line

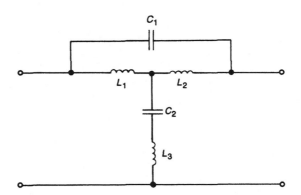

Fig. 10.9 Group-delay equalizer

inductor has a reactance which is directly proportional to frequency and the reactance of a capacitor is inversely proportional to frequency. If a particular band of frequencies is to be transmitted by a filter, series and shunt impedances are required which reach a maximum or a minimum value at the centre of the frequency band. Such impedances can be obtained by the use of series and parallel resonant circuits.

The Prototype or Constant-k Filter

Low-pass Filter

A low-pass filter should be able to pass, with zero attenuation, all frequencies from zero up to a certain frequency which is known as the *cut-off frequency* f_c. At frequencies greater than the cut-off frequency the attenuation of the filter will increase with increase in frequency up to a very high value. The basic prototype (or constant-k) T and π low-pass filter circuits are shown in Figs 10.10(a) and (b) respectively. For both circuits the total series impedance is ωL and the total shunt impedance is $1/\omega C$. The term 'constant-k' is used to denote that the product of the series and shunt impedances is a constant at all frequencies.

Suppose a voltage source of variable frequency is applied across the input terminals of the filter. At low frequencies the reactance of the series inductor L is low and the reactance of the shunt capacitor C is high; at these frequencies, therefore, the inductance offers little opposition to the flow of current while the capacitance has zero shunting effect. Low-frequency signals are propagated through the filter without loss. As the frequency of the input signal is increased, the inductive reactance will increase and the capacitive reactance will fall until, at the cut-off frequency f_c, the attenuation of the filter suddenly increases. Thereafter, the attenuation of the filter rises rapidly with increase in frequency. The ideal attenuation–frequency characteristic of a constant-k low-pass filter is shown by Fig. 10.10(c). In practice, an inductor inevitably possesses some resistance and because of this the filter does introduce some attenuation into the passband; also, the attenuation does not rise so sharply at the cut-off frequency. The practical attenuation–frequency characteristic of a low-pass filter is shown by the broken line in Fig. 10.10(c).

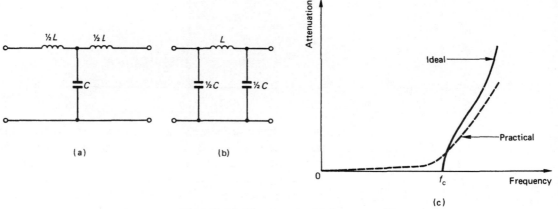

Fig. 10.10 The constant-k low-pass filter

High-pass Filter

The action of a high-pass filter is to transmit all those frequencies which are higher than its cut-off frequency and to prevent the passage of all lower frequencies. Figures 10.11(a) and (b) give the circuits of T and π constant-k high-pass filters. At low frequencies the series capacitance C has a high reactance and the shunt inductive reactance is low, so that low-frequency signals are attenuated as they travel through the filter. At high frequencies, on the other hand, the series reactance is low and the shunt reactance is high and the filter offers zero attenuation. The attenuation–frequency characteristic of the ideal high-pass filter is shown in Fig. 10.11(c), while the broken curve shows how the presence of resistance modifies the ideal characteristic.

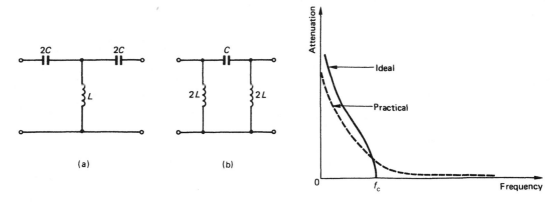

Fig. 10.11 The constant-*k* high-pass filter

Band-pass Filter

Figure 10.12(*a*) shows the circuit of a T constant-*k* band-pass filter and Fig. 10.12(*b*) shows its ideal and practical attenuation—frequency characteristics. Ideally, the filter passes, with zero attenuation, a particular band of frequencies and offers considerable attenuation to all frequencies outside this passband. The required characteristic is obtained by using two series-tuned circuits as the series impedance and a single parallel-tuned circuit as the shunt impedance. The three circuits are arranged to be resonant at the same frequency. For signals at or near this common resonant frequency, the series reactance is low and the shunt reactance is high so that the filter offers, ideally, zero attenuation. At frequencies either side of the required passband the tuned circuit's impedances have varied to such an extent that considerable attenuation is offered.

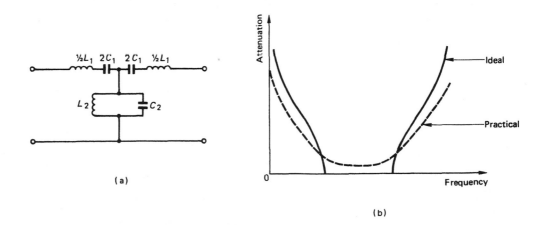

Fig. 10.12 The constant-*k* band-pass filter

Band-stop Filter

The fourth kind of filter, which is very much less often used, is the band-stop filter. This type of filter, shown in Fig. 10.13(*a*), provides a large attenuation to signals whose frequencies are within a particular frequency band. The ideal and the practical attenuation—frequency characteristics of a band-stop filter are shown in Fig. 10.13(*b*).

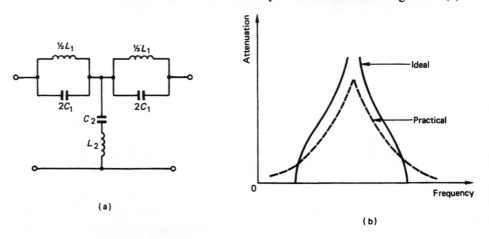

(a)

(b)

Fig. 10.13 The constant-*k* band-stop filter

Modern Filter Designs

In recent years many *LC* filters have been designed using techniques that allow a filter to be designed with a very accurate desired attenuation—frequency characteristic.

The attenuation—frequency characteristic of a Butterworth low-pass filter is shown in Fig. 10.14(*a*); it is maximally flat in the passband and has an attenuation of 3 dB at the cut-off frequency f_c. The Bessel

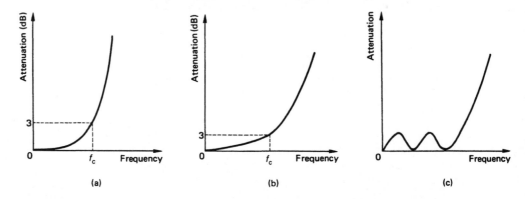

(a)   (b)   (c)

Fig. 10.14 Attenuation—frequency characteristics of (*a*) Butterworth, (*b*) Bessel and (*c*) Tchebyscheff low-pass filters

low-pass filter (Fig. 10.14(*b*)) introduces a constant time delay to all frequencies in the passband and it has 3 dB loss at the cut-off frequency. The attenuation of the Bessel filter does not rise as rapidly as that of the Butterworth filter. A more rapid increase in attenuation outside the passband can be obtained by the use of a Tchebyscheff filter (Fig. 10.14(*c*)) at the expense, however, of ripple in the passband.

Crystal Filters

For some applications the maximum selectivity that a band-pass *LC* filter can provide is inadequate, and in such cases a crystal filter can be employed. A crystal filter is one in which the required series and shunt impedances are provided by *piezoelectric crystals*.

Piezoelectric Crystals

A piezoelectric crystal is a material, such as quartz, having the property that, if subjected to a mechanical stress, a potential difference is developed across it, and if the stress is reversed a p.d. of opposite polarity is developed. Conversely, the application of a potential difference to a piezoelectric crystal causes the crystal to be stressed in a direction depending on the polarity of the applied voltage.

In its natural state, quartz crystal is of hexagonal cross-section with pointed ends. If a small, thin plate is cut from a crystal the plate will have a particular natural frequency, and if an a.c. voltage at its natural frequency is applied across it, the plate will vibrate vigorously. The natural frequency of a crystal plate depends upon its dimensions, the mode of vibration and its original position or *cut* in the crystal. The important characteristics of a particular cut are its natural frequency and its temperature coefficient; one cut, the *GT cut*, has a negligible temperature coefficient over a temperature range from 0 °C to 100 °C; another cut, the *AT cut*, has a temperature coefficient that varies from about +10 ppm/°C at 0 °C to 0 ppm/°C at 40 °C and about +20 ppm/°C at 90 °C. Crystal plates are available with fundamental natural frequencies from 4 kHz up to about 10 MHz or so. For higher frequencies the required plate thickness is very small and the plate is fragile; however, a crystal can be operated at a harmonic of its fundamental frequency and such *overtone* operation raises the possible upper frequency to about 100 MHz.

The electrical equivalent circuit of a crystal is shown in Fig. 10.15. The inductance *L* represents the inertia of the mass of the crystal plate when it is vibrating; the capacitance C_1 represents the reciprocal of the stiffness of the crystal plate; and the resistance *R* represents the frictional losses of the vibrating plate. The capacitance C_2 is the actual capacitance of the crystal (a piezoelectric crystal is an electrical insulator that is mounted between two conducting plates).

A series—parallel circuit, such as the one shown in Fig. 10.15 has

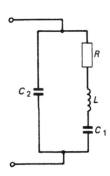

Fig. 10.15 Electrical equivalent circuit of a piezoelectric crystal

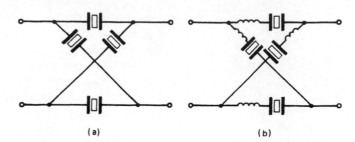

Fig. 10.16 Crystal filters

(a)   (b)

two resonant frequencies: the resonant frequency of the series arm RLC_1, and the parallel resonance produced by C_2 and the effective inductance of the series arm at a frequency above its (series) resonant frequency.

If a pair of similar crystals is connected in the series arms of a lattice network and another pair connected in the shunt arms, as shown in Fig. 10.16(a), the network will possess a band-pass characteristic, provided that the series resonant frequency of one pair is equal to the parallel-resonant frequency of the other pair. If the bandwidth is not wide enough it can be increased by the connection of an inductor of suitable value in series with the crystal (see Fig. 10.16(b)).

Active Filters

Inductors are relatively large and bulky components particularly at the lower frequencies, and they possess core and winding losses that are difficult to predict accurately and may vary with time, temperature and/or with frequency. The need for an inductor in a filter network can be avoided if a resistor—capacitor network is used as the feedback network of an amplifier. A number of different types of active filter are possible but the kind most commonly used since IC op-amps have become readily available is shown in Fig. 10.17. The circuit shown in Fig. 10.17(a) acts as a low-pass filter which can be given a Butterworth, a Bessel or a Tchebyscheff characteristic depending upon the values chosen for the various components. Changing over the positions of the resistors and the capacitors, as in Fig. 10.17(b),

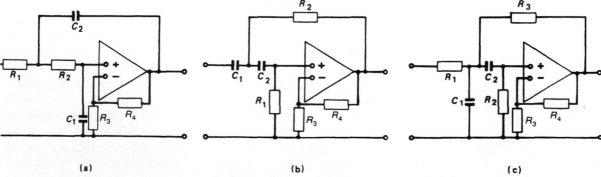

(a)   (b)   (c)

Fig. 10.17 Active filters: (a) low-pass, (b) high-pass and (c) band-pass

produces a high-pass filter with the required type of attenuation—frequency characteristic. Lastly, a band-pass characteristic is obtained by connecting the resistance—capacitance network in the manner shown by Fig. 10.17(*c*).

Exercises

10.1 Draw the circuit diagram of a basic low-pass filter and draw its (*a*) ideal and (*b*) practical attenuation—frequency characteristics. What is the general name given to this kind of filter?

10.2 A signal occupying the frequency band 50 Hz—25 kHz is applied to the input of the two cascaded filters shown in Fig. 10.18. Determine the bandwidth of the output signal.

10.3 If the cut-off frequencies of the two filters shown in Fig. 10.18 are switched over, what will then be the output bandwidth?

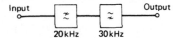

Fig. 10.18

10.4 The attentuation—frequency characteristic of a filter is given in Table 10.1. Plot the characteristic and state what kind of filter it is.

Table 10.1

Frequency (Hz)	10	30	100	300	1000	3000	10 000	30 000
Attenuation (dB)	30	20	10	1	1	2	20	30

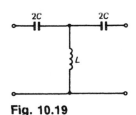

Fig. 10.19

10.5 (*a*) What kind of filter is shown in Fig. 10.19? (*b*) What is meant by the term 'constant-*k*'? (*c*) Briefly explain how the filter works. (*d*) Sketch a typical loss—frequency characteristic.

10.6 Figure 10.20 shows the attenuation—frequency characteristic of a filter, (*a*) What kind of filter is it? (*b*) What does f_c stand for? (*c*) Sketch the ideal characteristic of this filter.

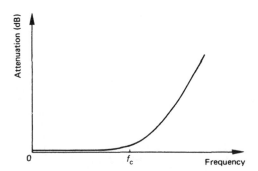

Fig. 10.20

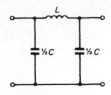

Fig. 10.21

Fig. 10.22

10.7 The low-pass filter shown in Fig. 10.21 is to have its loss just outside the passband increase more rapidly. Indicate how this can be done and say what the technique is called.

10.8 Determine the band of frequencies that appears at the output of Fig. 10.22.

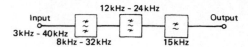

10.9 List the relative merits of *LC* and crystal filters.

10.10 Explain why there is a need for filters in communication networks.

10.11 A T attenuator is to have 9 dB loss and a characteristic resistance of 140 Ω. Design the circuit.

10.12 A π attenuator is to have 9 dB loss and a characteristic resistance of 140 Ω. Design the circuit.

10.13 The input signal to a line amplifier is at a level of −7 dBm. The amplifier has a gain of 27 dB and its output level is to be +10 dBm. Determine the loss of an attenuator to be placed at the input to the amplifier to achieve the requirements.

10.14 Design a T attenuator to have 15 dB loss and a characteristic resistance of 75 Ω.

10.15 Design a π attenuator to have 20 dB loss and a characteristic resistance of 600 Ω.

10.16 Draw the circuit of (*a*) an audio attenuation equalizer, (*b*) a low-pass filter. Explain how they differ from one another and describe the function of each.

10.17 The group delay—frequency characteristic of a circuit is described by the data given in Table 10.2. Write down the table that would describe the characteristics of the ideal group-delay equalizer for this circuit.

Table 10.2

Frequency (Hz)	400	800	2000	3000
Delay (ms)	0.01	0.005	0.07	0.30

11 The Access and Core Networks

The telephone network of the UK is divided into the *access network* and the *core network*. The access network comprises the local lines that connect the telephones, telephone switchboards, FAX machines, computers, etc. to the local telephone exchange. The core network consists of a large number of trunk circuits that interconnect telephone exchanges.

The core network part of the public switched telephone network (PSTN) consists essentially of two parts: one part provides switching to organize the required line connections; the other part provides the line network which interconnects telephone exchanges. Until the 1980s the PSTN was entirely based upon analogue switching with a predominantly analogue line network, although PCM systems were being introduced at a fast rate. Now the trunk network is entirely digital. Eventually, the access network will also be converted to digital operation but, because of the sheer size of the operation, this will take some time to complete.

Digital telephone exchanges are replacing analogue exchanges in the UK network at as fast a rate as possible with the intention of eventually converting the PSTN wholly to digital operation. By 1993 more than 70% of BT's customers were connected to a digital local exchange. The major reasons for the changeover from analogue to digital working are as follows.

(*a*) It is more economical.
(*b*) New services are more easily introduced.
(*c*) New electronic techniques (in LSI circuits) are mainly digital.
(*d*) Network management is more easily implemented.

The new digital network will be able to carry signals of all kinds with equal ease and is known as the integrated services digital network (ISDN). It is not economically possible to connect every exchange in the network directly to every other exchange; direct trunks are only provided between two exchanges when justified by the traffic carried. The remainder of the traffic is routed via the *integrated digital network* (IDN).

The IDN is a network in which all telephone exchanges are digital and all trunk circuits are carried by digital transmission systems. The

IDN includes some 5000 *remote concentrator units* (RCUs), about 400 *digital local telephone exchanges* (DLEs), and 53 fully interconnected digital telephone exchanges that are known as *digital main switching units* (DMSUs). The basic arrangement of the IDN is shown by Fig. 11.1(*a*). Customers are connected via a RCU to a digital local exchange and then each digital exchange has trunks to two DMSUs. All the 53 DMSUs are fully interconnected with one another. The network is augmented by access to international gateways, to operator services and to network services such as the 0800 Freephone. A number of analogue local telephone exchanges are still present in the UK telephone network and these gain access to the IDN via analogue-to-digital/digital-to-analogue converter units as shown by Fig. 11.1(*b*).

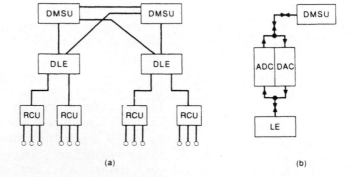

Fig. 11.1 (*a*) Integrated digital network, (*b*) access to IDN for analogue exchange
(RCU = remote concentrator unit, DLE = digital local exchange, DMSU = digital main switching unit)

The Access Network

The local line connection between a customer and the local RCU consists of a pair of copper conductors in a telephone cable. Since a large telephone exchange may have up to 10 000 customers the access network can be quite complicated, particularly because provision must be made for fluctuating demand. The access network is provided on the basis of forecasts made of the future demand for telephone service, the object being to provide service on demand as economically as possible. Since the demand for telephone service fluctuates considerably there is the problem of forecasting requirements and deciding how much plant should be provided initially and how much at future dates. No matter how carefully the forecasting is carried out, some errors always occur and allowance for this must be made in the planning and provision of cables, i.e. the access network must be flexible. An access network must be laid out so that the situation does not arise where potential customers cannot be given service in some parts of a telephone exchange area while in other parts of the same area spare cable pairs remain. The modern way of laying out the access network is shown in Fig. 11.2. Each customer's telephone is connected to a distribution point, such as a terminal block on a pole or a wall. The distribution points are connected by small distribution cables to secondary cross-connection points (SCPs), an SCP being a street structure that provides flexibility because it allows

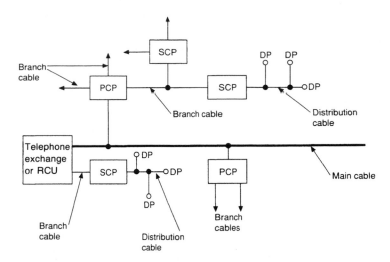

Fig. 11.2 Access network
(DP = distribution point,
PCP = primary cross-connection
point,
SCP = secondary cross-connection
point)

any incoming pair to be connected to any outgoing pair. The SCPs are connected by larger branch cables to primary cross-connection points (PCPs); these have the same function as SCPs, but are larger. Finally, main cables connect the PCPs to the telephone exchange.

The Core Network

Trunk circuits are employed to interconnect telephone exchanges and they are all routed over multi-channel PCM systems which terminate directly on digital switches in each telephone exchange.

Two-wire circuits are employed only for local lines in the access network, since all trunk circuits are routed over multi-channel PCM systems and these employ separate channels for each direction of transmission. The trunk circuits are therefore said to be four-wire operated, as before the advent of multi-channel systems four conductors were actually used for each trunk circuit. This means that if a connection set up via the PSTN involves the use of a trunk circuit, it is necessary to convert the two-wire local line into a four-wire trunk circuit at the local telephone exchange. The basic set-ups of a connection (*a*) between two telephones connected to the same exchange and (*b*) between two telephones on exchanges that are some distance apart are shown by Fig. 11.3(*a*) and (*b*).

In Fig. 11.3(*a*) both telephones are connected by a two-wire local line to the same local telephone exchange and these lines are connected together by the exchange switching equipment. When a call is made to a number on a distant exchange, Fig. 11.3(*b*), the telephone is connected to the local exchange by a two-wire line in the access network. At the exchange the call is switched to a trunk circuit that connects the exchange to either the destination exchange or to a trunk switching exchange. Thus trunk circuit will be routed over a PCM system. Since each channel in a PCM system is unidirectional, two channels are used, one for each direction of transmission. The channels

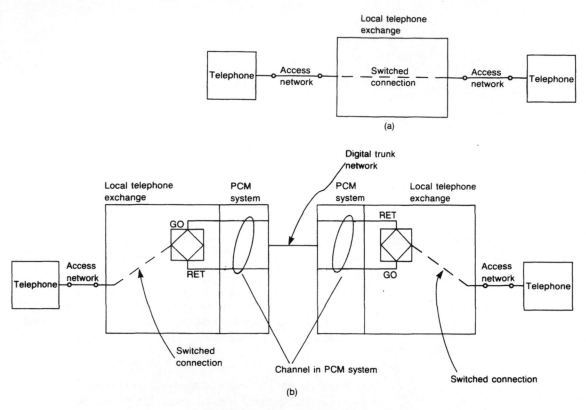

Fig. 11.3 Telephone connection between (a) two nearby local exchanges, (b) via the core network

are known as GO and RETURN. It is therefore necessary to convert a telephone circuit from two-wire operation in the access network to four-wire working in the core network and this conversion is the function of a *two-wire to four-wire converter*. At the destination telephone exchange the circuit must be converted back to two-wire operation and so another two-wire to four-wire converter must be employed for this purpose.

Two-wire to Four-wire Converters

A two-wire to four-wire converter is shown in Fig. 11.4. The operation of the converter is as follows: assume that, as is usually the case, the impedances of the circuits connected across terminals 1−1 and 2−2 are equal. A signal applied across the two-wire terminals of the unit will cause a current to flow and this current will induce an e.m.f. into each of the windings connected across terminals 1−1 and 2−2. The currents in these circuits are of equal magnitude and so they induce equal e.m.f.s into the balance circuit. The current flowing in the balance impedance is therefore zero. The power contained in the signal applied to the two-wire terminals is divided equally between terminals 1−1 and 2−2 so that the loss between these

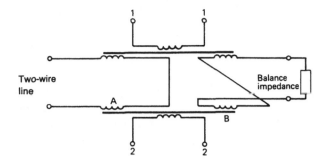

Fig. 11.4 A two-wire to four-wire converter

terminals and the input is 3 dB. In addition, transformer losses of about 1 dB are also present.

When a signal is applied to the terminals 2−2 of the circuit, the current which flows induces e.m.f.s, with the same polarity, into both winding A and winding B. Currents then flow in the two-wire and balance circuits that induce e.m.f.s of opposite polarity into the winding connected across terminals 1−1. If the balance impedance is adjusted to be equal to the impedance of the two-wire line, these induced e.m.f.s will be equal and they will cancel. Zero current will then flow at terminals 1−1. The signal power applied to terminals 2−2 is divided between the two-wire and balance circuits, giving a total loss of 4 dB between terminals 2−2 and the two-wire terminals. The loss between the terminals 2−2 and 1−1 is very high since little, if any, current flows at terminals 1−1. An alternative circuit for a two-wire to four-wire converter is shown in Fig. 11.5.

The operation of the converter depends on the accuracy with which the balance impedance simulates the impedance−frequency characteristic of the two-wire line. If the balance impedance does not exactly simulate the line impedance at all frequencies, the loss of the converter between the terminals 2−2 and 1−1 will be reduced and some energy will pass from terminals 2−2 to terminals 1−1.

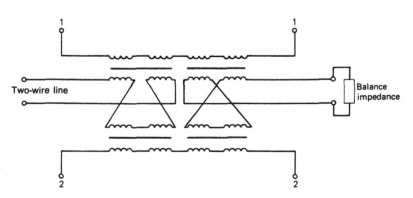

Fig. 11.5 Another two-wire to four-wire converter

Op-amp Two-wire to Four-wire Converter

A modem may employ an op-amp two-wire to four-wire converter to connect it to a two-wire presented telephone line. The circuit of such a converter is shown by Fig. 11.6. Op-amp 1 amplifies the differential output of the modem and this voltage is applied (i) to the primary winding of the transformer, (ii) to the inverting terminal of op-amp 2, and (iii) to the inverting terminal of op-amp 3. The voltage applied across the primary winding of the transformer is the sum of the output voltages of op-amps 1 and 2 and this voltage is applied via the transformer to the two-wire presented telephone line.

It is necessary to prevent the voltage transmitted to line also appearing at the input terminals of the modem and to achieve this the potential divider R_8 and R_9 applies a suitable voltage to the non-inverting terminal of op-amp 3. The voltage applied to this terminal is

$$V_3^+ = V_0 R_9/(R_8+R_9) + V_0^- R_8/(R_8+R_9)$$
$$= V_0^+ (R_9-R_8)/(R_8+R_9)$$

A suitable choice of the values of R_8 and R_9 will make the fed-back voltage equal to zero. Data signals received from the line are only amplified by op-amp 3 and are then applied to the modem input terminal.

The Interconnection of Trunk Circuits

The need often arises in the core network for trunk circuits to be connected in tandem in order to route a call from one exchange to

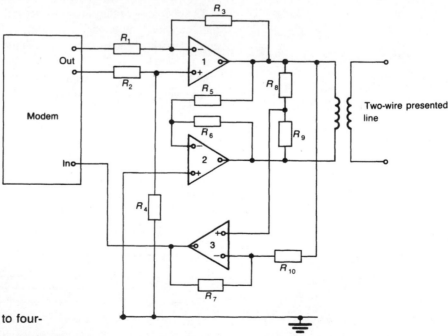

Fig. 11.6 Op-amp two-wire to four-wire converter

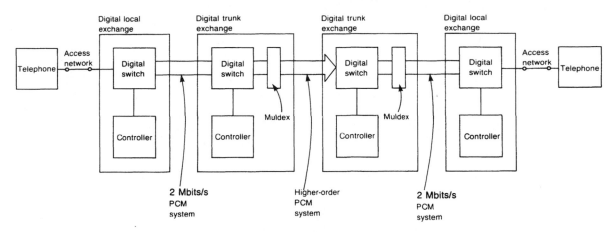

Fig. 11.7 Trunk circuits interconnect telephone exchanges

another. The trunk circuits linking digital telephone exchanges to the core network are routed over channels in a PCM multi-channel system and these channels each terminate directly on a digital switch, Fig. 11.7. Trunk circuits linking two digital local exchanges employ 30-channel 2 Mbits/s PCM systems but trunks between digital trunk exchanges, or DMSU, which carry more traffic, make use of higher-order TDM systems (page 183). Signalling between exchanges in the core network is carried out over a common signalling channel (usually channel 16 in a PCM system), that follows the ITU−T No. 7 recommendations. Each signalling system consists of two to four signalling channels connected in parallel, each of which follows a different physical path with automatic change-over to protect against traffic congestion and route failure. The basic arrangement for inter-exchange signalling is shown by Fig. 11.8. The ITU−T No. 7 system passes control information, such as the telephone number of the called telephone line and the telephone number of the calling telephone line, between two exchanges in the form of messages. Having the signalling circuits separated from the traffic channels allows control information to be transmitted at any time without causing any interference to speech circuits.

The IDN is a telephone-switching network in which all the telephone exchanges are digital types and all trunk circuits are routed over PCM systems. The basic layout of the IDN is shown in Fig. 11.9; each DLE has a direct link to its own DMSU and may also have direct connections to one or more DLEs. All the DMSUs are fully interconnected. When a telephone call is made from a telephone on one exchange to a telephone on another exchange the call may go via a direct link between the two exchanges. Nearly always the direct routes are between nearby exchanges because only then will there be enough telephone traffic to economically justify the provision of the direct route. In all other cases the call is first routed from the DLE to the DMSU; here it may be switched to another DLE or to

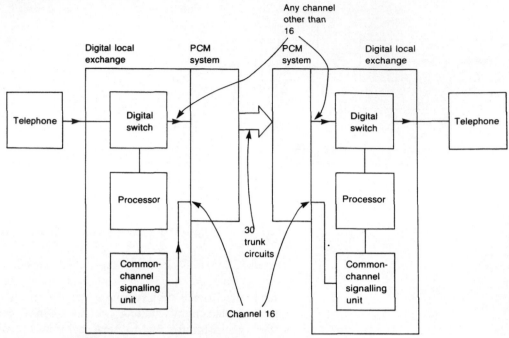

Fig. 11.8 Inter-exchange signalling

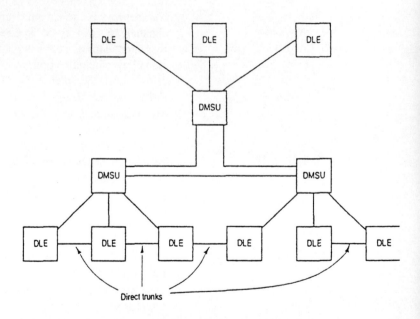

Fig. 11.9 The integrated digital network

another DMSU and from there to the destination DLE. In a DMSU the action is essentially to switch a channel in one PCM system to another channel in another PCM system.

The circuits between DLEs, between DLE and DMSU, and between two DMSUs are routed over multi-channel PCM systems. The

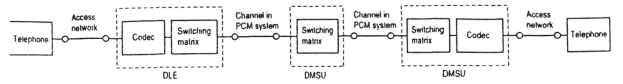

Fig. 11.10 A typical connection in the IDN

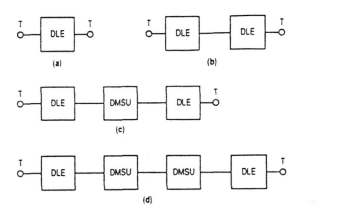

Fig. 11.11 Possible IDN connections

capacity of the basic PCM system is 30 telephone channels and the multiplexing process — i.e. grouping several telephone channels to work over a single transmission path — is carried out as part of the digital exchange switching action. The set-up of a typical call is shown in Fig. 11.10. Some of the possible connections that might be set up in the telephone network are shown in Fig. 11.11.

International Circuits

International direct dialling (IDD) is available from the UK to more than 200 other countries; this service allows any customer of BT to dial a telephone number directly to set up a connection with a telephone in another country. There is an international ISDN service that allows users to send or receive voice, text and image signals over a single digital link to another country. International private circuits (IPCs) are also provided on request; these can provide bit rates varying from 2400 bits/s to 2 Mbits/s. There is also an international FAX service.

Figure 11.12 shows the network which is used to provide IDD. Five fully connected *digital international switching centres* (DISCs), located at either London or Madely, are linked to the digital main switching units (DMSU) of the PSTN and also to a digital circuit multiplication equipment (DCME). A DCME uses voice compression techniques to increase the utilization of each international circuit and it is connected, by the *backhaul network*, to a number of *frontier stations*. Each frontier station gives access to either a submarine cable system or to the earth station of a communication satellite system. A call originated from a local telephone is routed via the PSTN to a DISC and thence to the appropriate DCME. Here the circuit is

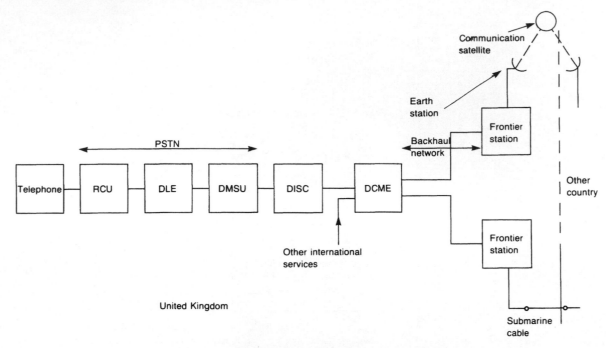

Fig. 11.12 International direct dialling circuit

multiplexed with the other international services before it is transmitted over the backhaul network to a frontier station. At the frontier station the circuit is connected to an international link over either a submarine cable system or a communication satellite system. At the other end of the international link the circuit is set up via the telephone network of the destination country.

Echoes in Telephone Circuits

An echo may appear in a telephone circuit if a signal is reflected from one direction of transmission to the other either at a two-wire to four-wire converter, or by acoustic coupling at either handset. The presence of an echo on a telephone line can make it difficult for a speaker to talk normally and it reduces intelligibility for the listener. Conversation becomes very difficult if the echo is fairly loud and the delay between the original signal and echo is more than about 100 ms. Echo is not audible if the delay is less than about 20 ms but merely gives the impression of an increase in sidetone. Echo starts to become audible when the delay exceeds about 30 ms. This means that echo only becomes a problem on very long telephone lines such as those used for international calls.

The effects of echoes on long lines can be removed by the use of either *echo suppression* or *echo cancellation*. An echo suppressor works by detecting when the distant speaker is talking and the near

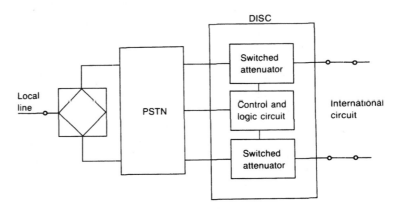

Fig. 11.13 Echo suppression

speaker is quiet. Any signal from the near speaker is then greatly attenuated by switching an attenuator into the circuit to prevent any echo occurring. The basic idea is illustrated by Fig. 11.13. An echo suppressor must be located near the source of the echo and so a suppressor is required at each end of the circuit. This type of echo suppression is effective but it makes it difficult for one person to attempt to speak while the other one is already talking. With echo cancellation a replica of the echo signal is synthesized by a complex circuit, known as an *adaptive filter*, and this replica signal is subtracted from the transmitted signal. The circuitry involved is very complex but is now available in a VLSI package. Figure 11.14 shows the location of an echo canceller in a long-distance trunk circuit.

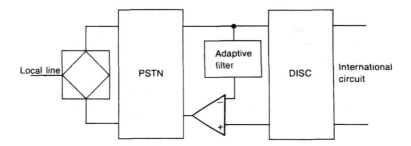

Fig. 11.14 Echo cancellation

HF International Radio Links

Two telephones in different countries may be connected together by a four-wire radio link operating in the high-frequency band 3–30 MHz. Figure 11.15 shows a typical arrangement.

A form of SSB working, known as independent sideband (ISB), enables four commercial-quality 300–3000 Hz bandwidth speech circuits to be accommodated in a 12 kHz bandwidth. A telephone in one country is connected, via the trunk network of that country, to an international telephone exchange. Here the call is established via a radio circuit to the required country and the telephone network in

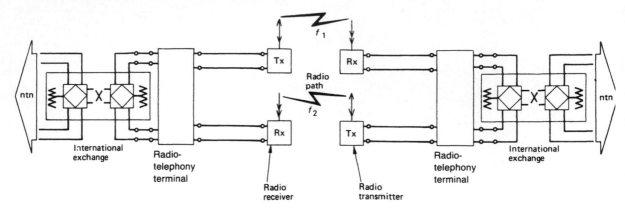

Fig. 11.15 An international radio-telephony circuit (ntn = national telephone network)

that country. Different frequencies are used for the two directions of transmission over the radio link to eliminate the possibility of oscillations occurring in the transmitter–receiver transmitter–receiver loop.

Ship Radio-telephones

Telephonic communication is often required between a telephone subscriber and a ship at sea and the arrangement for setting up such a connection is shown in Fig. 11.16. The telephone subscriber is connected, via the trunk network, with the control centre. The control centre is linked by cable to a number of coastal transmitting and receiving radio stations, each of which transmits to or receives from a different part of the world. The control centre establishes the required connection with a distant ship via the appropriate pair of radio stations. To provide good coverage of a particular area of the world each transmitting station transmits on several different frequencies at the same time. The actual frequencies depend upon the part of the world concerned and are given in Chapter 5.

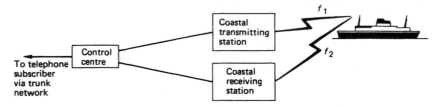

Fig. 11.16 Shore-to-ship telephone connection

INMARSAT

These MF and HF services are subject to adverse propagation conditions, poor quality reception and delays in obtaining service. To overcome these disadvantages a maritime satellite system has been

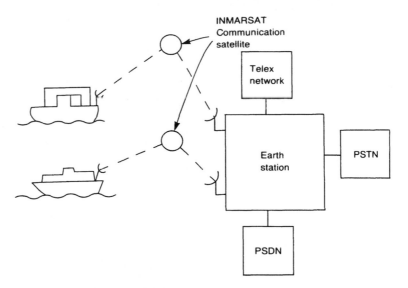

Fig. 11.17 INMARSAT service

introduced that provides a good quality service for the transmission of messages, but not at the present for real-time communications. The service, known as INMARSAT, is outlined by Fig. 11.17; it uses four satellites in the geostationary orbit to give earth-wide coverage except for the polar regions. Each satellite region is served by one *network co-ordination station* (NCS) which controls the operations within that region. A *land earth station* (LES) is the link between the *mobile earth stations* (MES) and the land networks of a country such as its PSTN. A MES is mounted on board a ship, or on a lorry, bus or train which can also have service. A MES must log in with a NCS before it is able to send or receive any traffic so that the location of all the MES in the system is known. Public access to the INMARSAT system is only possible via the Telex network; all other customers must register beforehand and are then given a PIN.

There are now five versions of INMARSAT in operation. INMARSAT A is the original system that was designed for use by ships at sea but is now also used by land-based customers. It can provide voice, FAX, data and Telex services; the voice channels use frequency modulation with a channel spacing of 50 kHz. INMARSAT B is a newer system that was intended as a replacement for INMARSAT A; it provides the same services but uses digital techniques for voice transmissions. INMARSAT C provides Telex, data, FAX and store and forward services, but no speech facility. With the store and forward service a message is transmitted to the land control station where it is stored and then re-transmitted to its destination. The portable system is small enough to fit inside a briefcase. The INMARSAT C service is used by both land and maritime customers. INMARSAT M provides voice, data and FAX services and is intended for use by small boats, but is can also be used on land. The mobile terminal is small enough to be carried in a briefcase. An airborne version of INMARSAT is also available; it provides voice, data and FAX facilities to customers flying in an aircraft.

Exercises

11.1 Draw the block diagram of a four-wire circuit that is routed over a PCM system and briefly explain its operation.

11.2 Discuss the reasons why a four-wire circuit may have an echo. How is such echoing overcome?

11.3 What is meant by *noise* in a transmission system? Give four sources of noise. Suggest how the effects of each source may be minimized

11.4 Draw the circuit of a two-wire to four-wire converter and explain its operation. Determine its loss (*a*) from two-wire input to GO output, (*b*) from RETURN input to GO output.

11.5 Draw the layout of a typical local telephone exchange area. Make clear the reasons why PCPs and SCPs are used.

11.6 Draw, and explain, an international radio-telephony circuit that uses a carrier in the high-frequency band. State the bandwidth provided by the system.
  Discuss the advantages and disadvantages of such a system. Why is it less popular now than several years ago?

11.7 What are the access and the core networks? What kinds of cable are predominantly used in each network? Why do all new trunk cables use optical fibre cable?

11.8 A telephone call is made from a number in Devon to a number in Northumberland. Draw a block diagram of a possible routing through the access and core networks. Why would there not be a direct circuit between the two local telephone exchanges?

12 Pulse-code Modulation

The basic principles of a time-division multiplex (TDM) system were outlined in Chapter 3, where it is shown how TDM allows a number of different channels to have access to the common transmission path for a short period of time. The methods of pulse modulation outlined in that chapter are, in practice, rarely employed since a much better performance can be achieved by the use of *pulse-code modulation* (PCM). However, the first stage in the production of a PCM signal employs pulse-amplitude modulation (PAM).

The ITU−T 30-channel PCM system* provides trunk circuits between two digital local exchanges (DLE). Higher-capacity systems, formed from a combination of 30-channel systems, are used to provide trunk circuits between two digital main switching units (DMSU) and between a DLE and a DMSU.

Pulse-amplitude Modulation

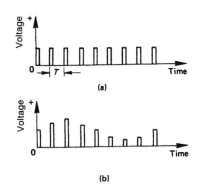

Fig. 12.1 Pulse-amplitude modulation: (a) unmodulated pulses, (b) PAM wave

* Often known as the CEPT (see p. 68) PCM system since its parameters and performance were first specified in Europe.

A time-division multiplex system is based upon the sampling of the amplitude of the information signal at regular intervals, and the subsequent transmission of one or more pulses to represent each sample. In an *analogue* pulse system, for the intelligence contained in the information signal to be transmitted the characteristics of the pulse must, in some way, be varied in accordance with the amplitude of the sample. The pulse characteristic which is varied can be the amplitude, or the width, or the position of the pulse, to give either pulse-amplitude, pulse-duration, or pulse-position modulation. For a *digital* pulse system, such as PCM, information about each sample is transmitted to the line by a train of pulses which indicate, using the binary code, the amplitude of that sample. The analogue methods of pulse modulation are rarely used in their own right in modern communication systems but pulse-amplitude modulation (PAM) is employed as a step in the production of a PCM signal.

With PAM, pulses of equal width and spacing have their amplitudes varied in accordance with the characteristics of a modulating signal. Figure 12.1(a) shows an unmodulated pulse train, often known as the *clock*, which has a periodic time of T seconds. Thus the number of pulses occurring per second − known as the pulse repetition frequency − is equal to $1/T$. The clock and the modulating signal

159

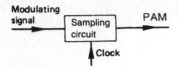

Fig. 12.2 Production of a PAM signal

are applied to the two inputs of a sampling circuit. The sampler produces an output signal, equal to the instantaneous value of the modulating signal, only when a clock pulse is present, see Fig. 12.2. Thus the PAM output of the sampler consists of successive samples of the modulating signal and, assuming a sinusoidal modulating signal, is shown in Fig. 12.1(b).

The PAM waveform contains components at a number of different frequencies. These are:

(a) The modulating signal frequency.
(b) The pulse repetition frequency and upper and lower side-frequencies centred about the pulse repetition frequency.
(c) Harmonics of the pulse repetition frequency and upper and lower sidefrequencies centred upon each of these harmonics.
(d) A d.c. component whose voltage is equal to the mean value of the PAM waveform.

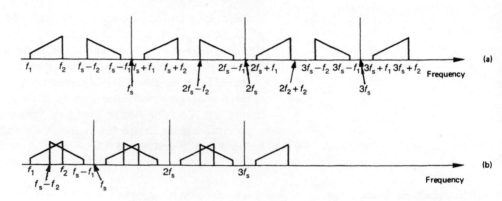

Fig. 12.3 Spectrum diagrams of a PAM signal: (a) sampling frequency f_s greater than twice the highest modulating frequency, (b) sampling frequency f_s less than twice the highest modulating frequency

The spectrum diagram of a PAM waveform is shown in Fig. 12.3(a). The modulating signal and each of the sidebands are represented by truncated triangles in which the vertical ordinates are made proportional to the modulating frequency, and no account is taken of amplitude. This method of representing signals shows immediately which sideband is erect and which is inverted. The modulating signal occupies the frequency band $f_2 - f_1$ and so the upper and lower sidebands of the pulse repetition frequency are:

$$(f_s + f_2) - (f_s + f_1) \quad \text{and} \quad (f_s - f_1) - (f_s - f_2) \quad (12.1)$$

The frequency spectrum of a PAM waveform contains the *original modulating signal* and this means that demodulation can be achieved by passing the PAM waveform through a low-pass filter as shown by Fig. 12.4. The low-pass filter must be able to pass the highest frequency f_2 in the modulating signal but it must *not* pass the lowest

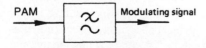

Fig. 12.4 Demodulation of a PAM waveform

frequency $f_s - f_2$ in the lower sideband of the pulse repetition frequency. Clearly, for this to be possible the lower sideband must not overlap the modulating signal in the way shown by Fig. 12.3(*b*). This means that the pulse repetition frequency, also known as the *sampling frequency* f_s, must be *at least twice* the highest frequency contained in the modulating signal; this requirement is known as the *sampling theorem*. In practice, the sampling frequency must be somewhat higher than twice the highest modulating signal frequency in order to provide a frequency gap in which the filter can build up its attenuation. For example, the ITU−T 30-channel PCM system uses an 8000 Hz sampling frequency for a highest modulating signal frequency of 3400 Hz.

If the sampling frequency f_s is not equal to at least twice the highest frequency in the signal an effect known as *aliasing* will occur. Suppose that the frequency of the input signal is 4.6 kHz. Then the sidefrequencies of the sampling frequency will be 3.4 kHz and 12.6 kHz and this means that the lower sidefrequency is lower than the signal frequency. The PAM waveform will then be the same as would be produced if the input signal were at 3.4 kHz signal, and this will be the frequency of the signal at the output of the demodulating low-pass filter. It is necessary, therefore, for the frequency of the input signal to a PCM system to be band-limited by a low-pass filter to a maximum of $f_s/2$ Hz.

Pulse-code Modulation

In pulse-code modulation (PCM) the total audio-frequency amplitude range to be transmitted by the system is divided into a number of allowable voltage levels, known as *sampling levels* or *quantum levels*. Each of these quantum levels is allocated a number as shown by Fig. 12.5 in which eight levels have been drawn. The analogue signal is sampled at regular intervals to produce a pulse-amplitude modulated waveform, but then the pulse amplitudes are rounded off to the nearest

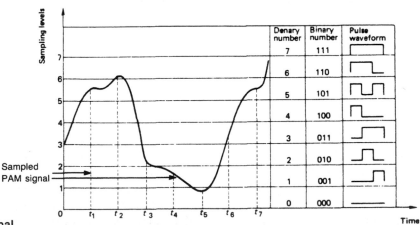

Fig. 12.5 Quantization of a signal

allowed quantum level. The process of approximating the sampled signal amplitudes to the nearest quantum level is known as *quantization*. The number of the nearest quantization level to each sampled amplitude is encoded in the equivalent binary form and is then transmitted to the line. Usually, a pulse is sent to represent binary 1 and no pulse is sent to represent binary 0.

Consider Fig. 12.5. The signal waveform is sampled at time intervals t_1, t_2, t_3, etc. At time t_1 the instantaneous signal amplitude is between levels 5 and 6 but, since it is nearer to level 6, it is approximated to this level. At instant t_2 the signal voltage is slightly greater than level 6 but is again rounded off to that level. Similarly, the sample taken at time t_3 is represented by level 2, the t_5 sample by level 1, and so on. Each quantized sample is then encoded into the binary-coded pulse waveform shown alongside. The binary pulse train which would represent this signal is shown in Fig. 12.6. A space, equal in time duration to one bit, has been left between each binary number.

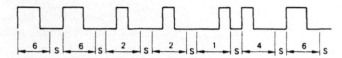

Fig. 12.6 Binary pulse train representing the signal shown in Fig. 12.5

Since the signal information is transmitted in binary form, the number of quantum levels used is always some power of 2, starting from zero; thus the highest numbered level is $2^n - 1$, where n is the number of bits used to represent each sample. In Fig. 12.5, 2^3 or eight levels only were drawn for clarity but practical systems employ many more quantization levels. For example, the ITU−T 30-channel system has 256 levels so that $n = 8$.

The signal received at the far end of the system is not an exact replica of the transmitted signal but, instead, is a quantized approximation to it. Fig. 12.7 shows a particular modulating signal and its quantized approximation. The difference between the two is an error waveform which, since it sounds noisy at the output of the system, is known as *quantization noise*. The magnitude of the quantization error depends upon the number of quantization levels used *and* the sampling frequency; increase in either or both of these parameters produces a reduction in the error. This is illustrated by Fig. 12.7 in which the effect of increasing the number of quantization levels from 8 to 16 and then to 32 is shown by (*a*), (*b*) and (*c*). Also, Figs 12.7(*b*) and (*d*) show the effect of an increase in the sampling frequency. The error waveforms for these four cases are given in Fig. 12.8.

Unfortunately, as will be shown later, increasing the number of quantization levels and/or increasing the sampling frequency results in a requirement for a wider bandwidth.

The output waveform of a PCM system can be regarded as consisting of the original modulating signal plus quantization noise. Quantization noise is only present at the output of a PCM system when a signal is being transmitted. The maximum quantization error is the same for all amplitudes of signal and so the signal-to-noise ratio at the output is worse for the smaller amplitude signals. Some improvement can be realized if the most significant bit is used to indicate whether the sampled voltage is positive or negative with respect to earth. Thus, if eight bits are used, as in the ITU−T 30-channel system, the quantization levels could be numbered in the manner shown in Fig. 12.9. If, for example, the sampled amplitude was rounded off to the twelfth positive quantum level the binary code would be 10001100, and if the sampled amplitude was represented by the eleventh negative quantum level the code would be 00001011.

Example 12.1

Calculate the percentage reduction in the quantization error, or noise, when the number of quantization levels is increased from 256 to 512.

Solution

The increase in the number of sampling levels is twofold. Hence the quantization error is reduced by one-half or 50%.   (*Ans.*)

Quantization Noise

The greater the number of quantum levels used the smaller will be the error involved in the quantization process and hence the smaller will be the quantization noise. The maximum quantization error is equal to one-half the height of a quantization step and this maximum error occurs when the signal voltage suddenly changes direction, see Fig. 12.10.

The r.m.s. value of a triangular waveform is equal to (maximum value)/$\sqrt{3}$ and hence the maximum r.m.s. error is $1/(2\sqrt{3}) \times$ height of quantum step.

The ratio

$$\text{(peak-to-peak signal voltage)/(r.m.s. noise)} = 2\sqrt{(3)}\, 2^n$$

and so the ratio

$$\text{(r.m.s. signal voltage)/(r.m.s. noise)} = \sqrt{(3/2)} \times 2^n \quad (12.2)$$

In decibels this is

$$\text{Signal-to-noise ratio} = 1.76 + 6n \text{ dB} \quad (12.3)$$

If n is greater than 4 the signal-to-noise ratio is approximately equal to 2^n and, since bandwidth is directly proportional to n, this means that the signal-to-noise ratio increases exponentially with increase in the bandwidth.

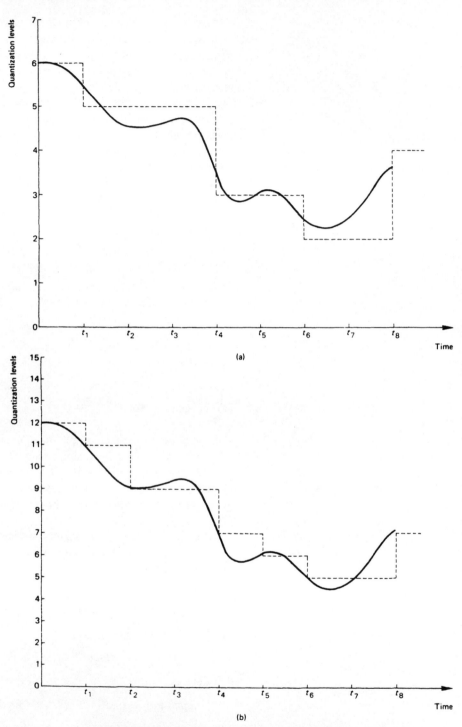

(a)

(b)

Fig. 12.7 Modulating signal and its quantized version (*a*), (*b*), and (*c*) showing the effect of increasing the number of quantization levels; (*b*) and (*d*) show the effect of increasing the sampling frequency

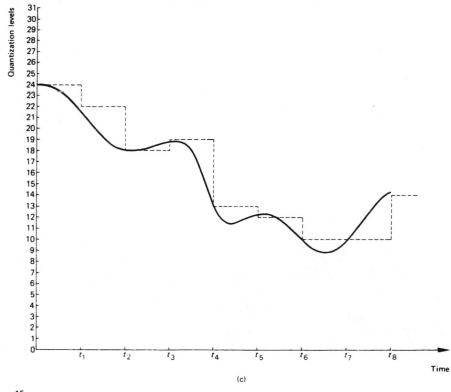

(c)

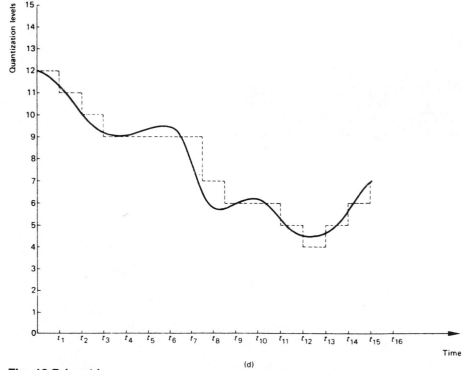

(d)

Fig. 12.7 (cont.)

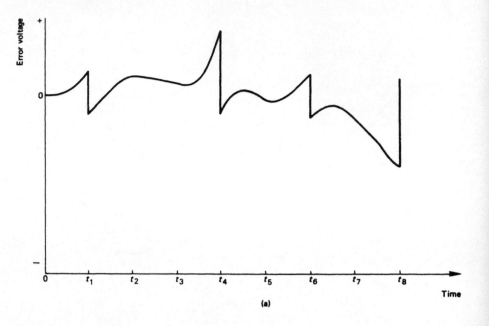

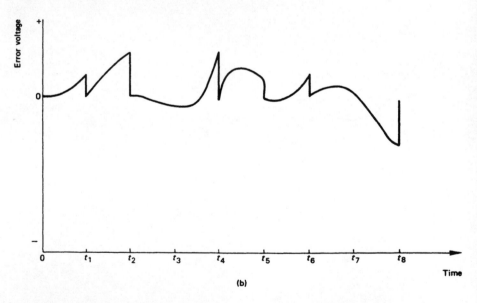

Fig. 12.8 Quantization error waveforms for Fig. 12.7

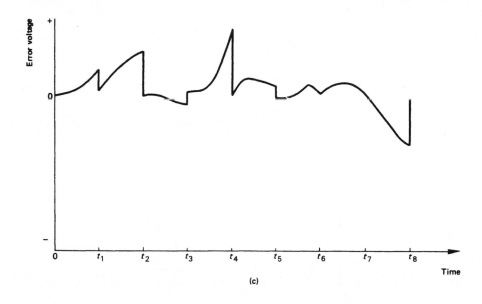

(c)

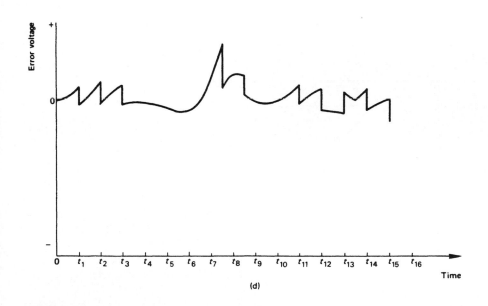

(d)

Fig. 12.8 (cont.)

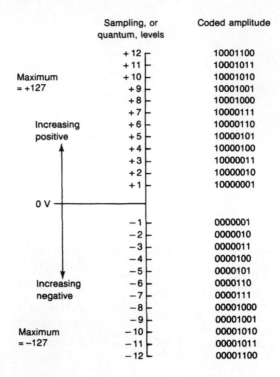

	Sampling, or quantum, levels	Coded amplitude
	+12	10001100
	+11	10001011
Maximum	+10	10001010
= +127	+9	10001001
	+8	10001000
	+7	10000111
Increasing	+6	10000110
positive	+5	10000101
	+4	10000100
	+3	10000011
	+2	10000010
	+1	10000001
0 V		
	−1	0000001
	−2	0000010
	−3	0000011
	−4	0000100
	−5	0000101
Increasing	−6	0000110
negative	−7	0000111
	−8	00001000
	−9	00001001
Maximum	−10	00001010
= −127	−11	00001011
	−12	00001100

Fig. 12.9 Sampling levels

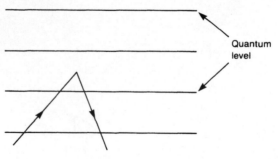

Fig. 12.10 Signal producing maximum quantization error

Quantum level

Example 12.2

Calculate the signal-to-noise ratio of a PCM system that uses 256 quantum levels.

Solution

$256 = 2^n$, $n = 8$.
Signal-to-noise ratio $= 1.76 + 6 \times 8 = 49.76\,\text{dB}$ *(Ans.)*

For the receiver at the end of a PCM channel to be reliably able to determine at each instant whether a pulse is, or is not, being received the bearer circuit must have a certain minimum signal-to-noise ratio. The effect of line noise is to cause an error in one bit; the significance of this depends upon which bit is affected, being least if the least significant bit is affected and greatest if the most significant bit is in error. It is found that if the signal-to-noise ratio is higher than 21 dB the bit error rate is fairly small, but the error rate increases rapidly if the signal-to-noise falls below 21 dB. Noise arising on the bearer circuit is additive to the quantization noise.

Non-linear Quantization

The use of equally spaced quantization levels as so far assumed is useful as long as the signal amplitude is large enough to cover several quantization levels. For small-amplitude speech signals, however, the output signal-to-noise ratio may well be inadequate. In the extreme case a very small-amplitude signal having a peak voltage insufficient to reach the first quantum level will not produce any change in the transmitted binary number — other than the sign digit as the signal crosses the zero voltage quantum level. This is shown by Fig. 12.11.

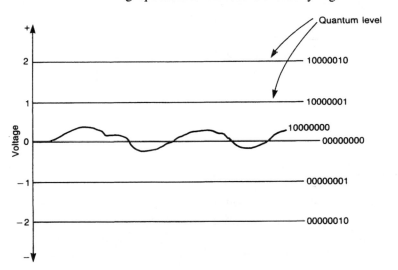

Fig. 12.11 Small-amplitude signal is not quantized

The signal-to-noise ratio for small signals can be improved, without increasing the number of quantization levels, by the use of non-linear quantization. Then the quantization levels are no longer equally spaced but, instead, are much closer together near the middle of the amplitude range than near the two ends of the range. This will ensure that small-amplitude signals are precisely quantized, while the larger-amplitude signals are quantized more coarsely. The spacing of the quantization steps is arranged to follow a logarithmic law such that a more or less constant output signal-to-noise ratio is obtained for signals of all allowable amplitudes.

The spacing of the quantum levels using non-linear quantization follows the ITU—T *A-law recommendations*. This law gives a segmented approximation to the two-part expression:

$$|V_{out}| = [1 + \log_e(A \ |V_{in}|)]/(1 + \log_e A) \qquad (12.4)$$

for $1/A \le |V_{in}| \le 1$

and

$$|V_{out}| = A \ |V_{in}|/(1 + \log_e A) \qquad (12.5)$$

for $0 \le |V_{in}| \le 1/A$

A is the *compression coefficient*; its value is 87.6.

The segmented approximation has each of eight successive segments changing their slope by a factor of 2, as shown by Fig. 12.12. The figure applies to both positive and negative voltages. Signal voltages from 0 to 1/128 of the maximum voltage are covered by the first 16 quantum levels, voltages from 1/128 to 1/64 times the maximum voltage are covered by the next 16 levels, and so on, with the last 16 levels covering voltages ranging from one-half the maximum value up to the maximum voltage.

Data signals are amplitude-limited to the range 0 to $1/A$.

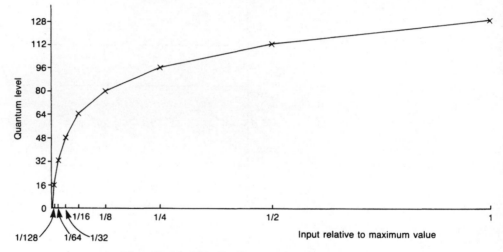

Fig. 12.12 ITU–T A-law quantization

The improvement in the quantization accuracy that non-linear quantization can give is illustrated by Fig. 12.13. The same small-voltage signal has been applied to a linear quantizer in Fig. 12.13(*a*) and to a non-linear quantizer in Fig. 12.13(*b*). It is clear that the quantization error is much smaller in the second case.

Non-linear quantization can be obtained in either one of two different ways:

(*a*) the analogue signal can have its range of amplitudes compressed and then encoding can take place, or
(*b*) a non-linear encoder can be employed.

Method (*b*) is employed in the ITU–T 32-channel system.

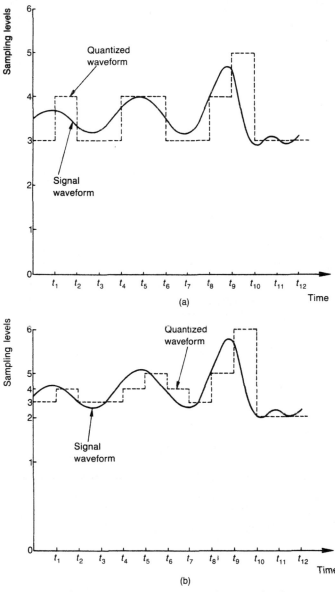

Fig. 12.13 Showing the improvement in quantization accuracy obtained by the use of non-linear quantization: (*a*) linear, (*b*) non-linear quantization

Time-division Multiplexing

The minimum sampling frequency, i.e. the number of samples taken per second, is equal to twice the highest frequency contained in the analogue signal. Thus, if the highest modulating signal is limited to 3400 Hz by an input low-pass filter, the minimum sampling frequency will be 6800 Hz. In practice, a sampling frequency somewhat greater than the minimum allowable would be used. The ITU−T 30-channel PCM system uses a 8000 Hz sampling frequency for the maximum audio frequency of 3400 Hz.

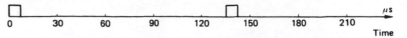

Fig. 12.14 Showing the time that a single channel PCM system occupies a transmission path

Each sample of the information signal is represented by eight bits, the first bit indicating the polarity of the sampled amplitude. Each bit is 488 ns wide and, hence, each sample occupies a time period of $8 \times 0.488 = 3.9\ \mu s$. A sample is taken every 1/8000 s or every $125\ \mu s$, and this means that the larger part of each sample period is unoccupied, see Fig. 12.14. The unused time periods can be used to carry other PCM channels; the number which can be fitted into the time available is $125/3.9 = 32$ channels. Two of these channels are used for synchronization and signalling purposes (that is, channels 0 and 16) and so 30 channels are available to carry speech signals.

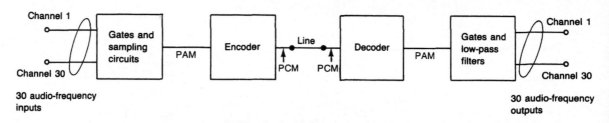

Fig. 12.15 Basic block diagram of 30-channel PCM system

The basic block diagram of the PCM system is shown by Fig. 12.15.

The channels are interleaved sequentially, sample by sample, and the time period occupied by 32 time slots is known as a *frame*. Thus a frame occupies $125\ \mu s$ and contains $32 \times 8 = 256$ bits. The line bit rate is

$$256 \times 8000 = 2048\ \text{kilobits/s (or 2.048 Mbits/s)}$$

This is usually referred to as 2 Mbits/s.

For a given signal-to-noise ratio, the use of non-linear quantization reduces the number of bits needed to represent each sample from 12 to 8.

Example 12.3

Calculate the percentage reduction in the bit rate per channel due to the use of non-linear quantization.

Solution
Bit rate = sampling frequency × bits per word = $8000 \times 12 = 96$ kbits/s
or $8000 \times 8 = 65$ kbits/s
Reduction = $(96 - 64)/96 \times 100 = 33.3\%$ (*Ans.*)

Bandwidth of a PCM Waveform

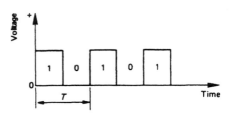

Fig. 12.16 PCM waveform having the highest fundamental frequency

The allowable amplitude range of a PCM system is divided into 2^n quantization levels. Each time the analogue signal is sampled, n bits are transmitted to indicate the appropriate quantum level. The analogue signal is sampled f_s times per second, where f_s is the sampling frequency, and so the number of bits transmitted per sample per channel is nf_s.

If the TDM system has m channels then the bit rate is

$$\text{Bit rate} = mnf_s \qquad (12.6)$$

i.e. bit rate = number of channels × sampling frequency × number of bits per word. The maximum fundamental frequency of the transmitted binary pulse train occurs when it consists of alternate 1s and 0s as shown in Fig. 12.16. The maximum fundamental frequency is the reciprocal of the periodic time of the PCM waveform and it is equal to one-half of the bit rate. The minimum frequency of the PCM waveform occurs when the binary number consists of consecutive 1 or 0 bits and is equal to 0 Hz. The minimum bandwidth which must be provided for a PCM waveform, assuming that only the fundamental frequency component need be transmitted and that the system is noise free, is thus from 0 Hz to (bit rate)/2 Hz.

$$\text{Minimum bandwidth} = (\text{bit rate})/2 \qquad (12.7)$$

Example 12.4

Calculate (*a*) the bit rate per channel, (*b*) the total bit rate and (*c*) the minimum bandwidth needed for the ITU−T 30-channel PCM system.

Solution
In this system 256 quantum levels are indicated by eight bits. Hence, since the sampling frequency is 8 kHz
(*a*) Bit rate per channel = 8 × 8000 = 64 kbits/s   (*Ans.*)
(*b*) Total bit rate   = 64 kbits/s × 32 = 2.048 Mbits/s   (*Ans.*)
(*c*) Minimum bandwidth = bit rate/2 = 1.024 MHz   (*Ans.*)

Frame Structure

The period of time occupied by the 32 channels is known as a *frame*. The frame structure of the 30-channel system is shown in Fig. 12.17. Sixteen frames form one *multi-frame* which occupies a time period of 2 ms. Frame 0 has its first four bits used for multi-frame alignment and some, or all, of its remaining bits used for alarm and supervisory purposes. Frames 1 to 15 are the signal frames and an exploded view of frame 7 is given. It can be seen to contain 32 time slots (0−31) each of which occupies a time period of 3.9 μs. Time slot 0 contains the frame synchronization signals and time slot 16 is (mainly) devoted to carrying telephone exchange signalling information. For frame 0, only the time slot 16 carries a multi-frame synchronization signal. Thus, 30 channels are made available for the transmission of speech. Synchronization is necessary to ensure that the information transmitted by, say, channel 1, is directed to channel 1 at the receiving end of the system.

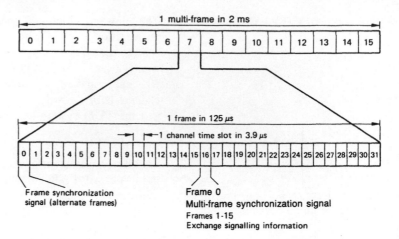

Fig. 12.17 30-channel PCM frame structure

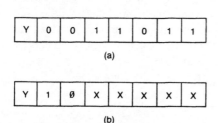

(a)

(b)

Fig. 12.18 Time slot 0 frame synchronization signal. (*a*) Frame alignment word, (*b*) contents in alternate frames.

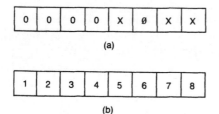

(a)

(b)

Fig. 12.19 Time slot 16 Multi-frame alignment signal (*a*) frame 0, (*b*) frames 1−15 signalling bits.

The structure of the time slot 0 frame synchronization signal, which occurs in alternate frames, is shown by Fig. 12.18(*a*) and the contents of the time slot in alternate frames are as shown by Fig. 12.18(*b*). The symbols used in the figures are:

Y: reserved for international use, normally at 1.
X: not allocated to any particular function, normally at 1.
Ø: normally 0 but is changed to 1 if a loss of frame alignment should occur.

The frame alignment pattern is one that will not occur very often when speech signals are being transmitted. Frame synchronization circuitry in the receiver checks the presence, or the absence, of the pattern in alternate frames. If the pattern is received correctly the receiver is able to synchronize with the transmitter. Once frame synchronization has been achieved multi-frame alignment is then obtained by the receiver searching for the word 0000 in the four most significant bits of time slot 16. When they are found frame 0 of the multi-frame is defined. The multi-frame alignment signal is shown in Fig. 12.19(*a*).

Time slot 16 carries a multi-frame synchronizing signal in frame 0, and telephone exchange signalling information in frames 1 to 15. In frame 0 the multi-frame synchronization signal has the form shown by Fig. 12.19(*b*). The first four bits form the alignment word. When the time slot is used for the transmission of signalling information in frames 1 to 15, the bits 1 to 4 are used for channels 1 to 15, and bits 5 to 8 are used for channels 16 to 30, in consecutive frames.

Multi-frame alignment is necessary to ensure that channel 1 at the transmitting end of a system is correctly taken as being channel 1 at the receiving terminal, and similarly for all the other 29 channels. Synchronization signals are also required to make sure that the clock in the receiving equipment runs at exactly the same speed as the transmitter clock. Any small differences would mean that there would not be a constant phase relationship between the timing of the two terminals and then every so often there would be a loss of frame alignment.

Signalling

The 64 kbits/s capacity of the signalling channel means that it is able to satisfy all the signalling requirements of the 30 speech channels with no interference with any speech or data transmissions.

Each speech channel is allocated four signalling bits in every 15th frame so that $30 \times 4 = 120$ bits are needed. This demands that 16 frames, giving $16 \times 8 = 128$ bits, are used. The first and last frames (0 and 15) carry frame synchronization bits; the other frames each carry the signalling bits of two adjacent channels. Thus frame 1 serves channels 1 and 17, frame 2 serves channels 2 and 18, and so on up to frame 15 which serves channels 15 and 31. This is shown by Fig. 12.20.

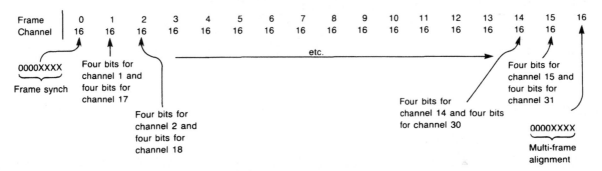

Fig. 12.20 Signalling in a PCM system

Line Signals

The information about the sampled amplitudes is signalled to the line using the binary code. So far it has been assumed that unipolar pulses are transmitted, i.e. a positive voltage representing binary 1 and zero voltage representing binary 0. This line code has the disadvantages that

(a) It always has a d.c. component, which makes it difficult to distinguish between signals and the d.c. power fed to the regenerators (for a coaxial cable system), and removal of the d.c. component will cause droop (p. 126) which results in drift relative to the decision threshold.

(b) It contains low-frequency components of relatively large amplitude which may be the source of crosstalk.

(c) It might contain a series of identical bits and this would make it difficult for the receiver or for a pulse regenerator to derive its clock signal from the incoming bit stream and the timing might be lost.

Alternate Mark Inversion

The disadvantages of unipolar signalling can be overcome by the use of *alternate mark inversion* (AMI), and the principle of this method of signalling is shown by Fig. 12.21. The basic unipolar signal (Fig. 12.21(*a*)) has alternate 1 bits inverted as shown by Fig. 12.21(*b*). This procedure ensures that each 1 bit is of opposite sign to the preceding 1 bit to make the d.c. component zero. Alternate mark inversion reduces the probability of a long stream of 0s being transmitted to line when a channel is idle.

AMI is an example of a *ternary* pulse waveform, i.e. it has three levels, positive, negative and zero.

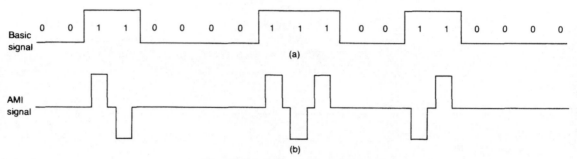

Fig. 12.21 Alternate mark inversion

High-density Bipolar Code

Alternate mark inversion signalling has the disadvantage that a long string of 0s in a bit stream results in no pulses being sent to line. The line pulse regenerators derive their timing information from the pulses incoming to them and so their operation is adversely affected by a period of no pulses. To overcome this problem a mark or 1 bit can be inserted into the bit stream after a given number of 0s have been sent. This kind of signalling is known as *high-density bipolar n* (HDB*n*). The code used with the ITU−T 30-channel PCM system,

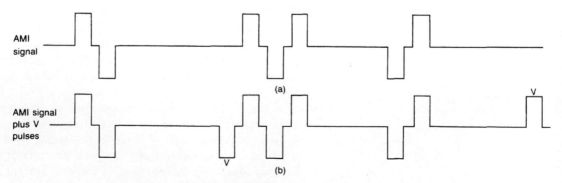

Fig. 12.22 HDB3 code: V pulse

and for the interface between 2/8 and 8/34 Mbits/s higher order systems, is HDB3, in which only three consecutive 0s are allowed to occur. If a fourth consecutive 0 arrives it is replaced by a 1 bit of the same polarity as the preceding 1. Hence this inserted 1 bit is of such polarity that it violates the AMI sequence; this is shown by Fig. 12.22.

This V pulse introduces a d.c. component to the signal and to remove this a B pulse of the opposite polarity is introduced three places before the violation. The basic rules for HDB3 coding are:

(a) All groups of four consecutive 0s are changed to 000V, where V is a violation pulse of the same polarity as the preceding 1.

(b) Successive violation bits must be of opposite polarity to eliminate the d.c. component.

(c) The number of bits between successive V bits must always be odd. If necessary, an extra B, or parity, bit must be added to satisfy this requirement. Whenever this occurs the V bit is still the violation pulse and the B bit must satisfy the AMI requirement. The B bit also ensures that consecutive V bits are of opposite polarity. An even number of 1s between two V bits would result in successive V bits having the same polarity. To prevent this from happening the first 0 of a four-0 bit sequence is changed to a 1 bit, called a parity bit, and the fourth 0 bit is changed to a 1 which violates the parity bit. This is shown by Fig. 12.23.

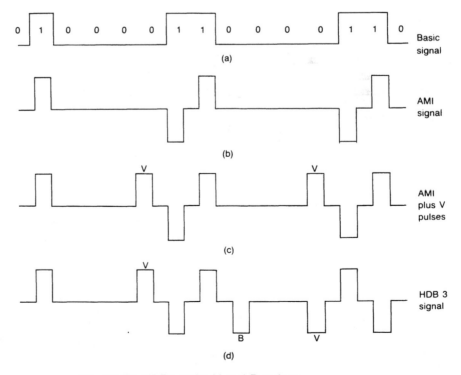

Fig. 12.23 HDB3 code: V and B pulses

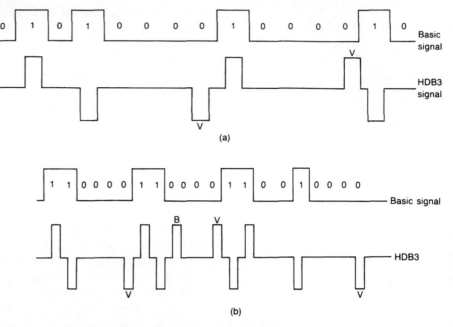

Fig. 12.24 Two examples of HDB3 code

(d) Signal bits that arrive after a V bit must also satisfy AMI and this means that they may need their polarity reversed.

(e) The B bit has the same polarity as the following V bit.

Two examples of the use of the HDB3 code are shown by Fig. 12.24(a) and (b).

Coded Mark Inversion

Coded mark inversion (CMI) is used for the interface between 140 Mbits/s equipment and for optical fibre systems. With CMI a 0 bit is represented by 01 and a 1 bit is alternately represented by 00 and 11, i.e. 0 by a half-width pulse and 1 by a full-width pulse with alternate marks being inverted, see Fig. 12.29. This is a form of 1B−2T code in which two symbols are used to convey one bit of information. Optical fibre cables have a much smaller attenuation than does coaxial cable and hence there is a reduced need for the line baud speed to be as small as possible. ON−OFF modulation is usually employed and this generates a binary line signal. Long-distance optical fibre PCM systems often employ the 5B−6T code in which six symbols convey five bits of information. There are then $2^6 = 64$ possible code words and 32 of these are used to represent 5-bit binary words.

A more complex code, known as 6B−4T, is used for 140 Mbits/s systems operating over coaxial cable. In this code four ternary symbols are used to represent six bits.

At 565 Mbits/s CMI cannot be used; then scrambled AMI is commonly employed.

Pulse Regenerators

During its transmission along a telephone line, the PCM signal is both attenuated and distorted but, provided the receiving equipment is able to determine whether a pulse is present or absent at any particular instant, no errors are introduced. To keep the pulse waveform within the accuracy required, *pulse regenerators* are fitted at intervals along the length of the line. The function of a pulse regenerator is to check the incoming pulse train at accurately timed intervals for the presence or absence of a pulse. Each time a pulse is detected, a new undistorted pulse is transmitted to line and, each time no pulse is detected, a pulse is not sent.

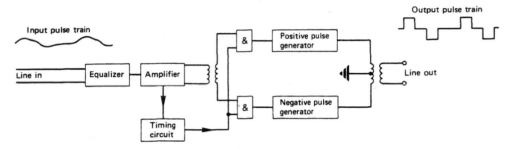

Fig. 12.25 Pulse regenerator

The simplified block diagram of a pulse regenerator is shown in Fig. 12.25. The incoming bit stream is first equalized and then amplified to reduce the effects of line attenuation and group delay—frequency distortion. The amplified signal is applied to a timing circuit which generates the required timing pulses. These timing pulses are applied to one of the inputs of two two-input AND gates, the phase-split amplified signal being applied to the other input terminals of the two gates. Whenever a timing pulse *and* a peak, positive or negative, of the incoming signal waveform occur at the same time, an output pulse is produced by the appropriate pulse generator. It is arranged that an output pulse will not occur unless the peak signal voltage is greater than some predetermined value in order to prevent false operation by noise peaks.

Provided the bit stream pulses are regenerated before the signal-to-noise ratio on the line has fallen to 21 dB, the effect of line noise on the error rate is extremely small. This means that impulse noise can be ignored and white noise is *not* cumulative along the length of the system. This feature is in marked contrast with an analogue system in which the signal-to-noise ratio must always progressively worsen towards the end of the system. Thus, the use of pulse regenerators allows very nearly distortion-free and noise-free transmission, regardless of the route taken by the circuit or its length. The main difficulty that arises is avoiding *jitter*. Jitter is caused by changes in phase in the regenerator which are mainly the result of noise and/or inadequate timing.

Many 2 Mbits/s PCM systems have been installed using existing audio-frequency cables; then the pulse regenerator spacing is 1.828 km because this was the spacing at which the loading coils were inserted into cable pairs. Newer 140 Mbits/s systems, which are routed over coaxial cable, have regenerators spaced at 2 km intervals. When a PCM system is routed over an optical fibre cable the line losses are so small that neither line pulse regenerators nor equalizers are needed unless the route length is in excess of about 40 km.

Pulse-code Modulation Systems

The basic block diagram of a 30-channel PCM system is shown by Fig. 12.26, in which 30 band-limited (to 4 kHz) channels are sampled sequentially at a sampling frequency of 8000 Hz to produce PAM waveforms. The PAM waveforms are interleaved on the common highway to produce a time-division multiplex signal before the multiplex signal is encoded by the non-linear encoder to give a unipolar PCM signal.

Synchronization and signalling bits are then added before the signal is converted to the HDB3 code and is then transmitted to the line at 2048 kbits/s. The transmitted bit stream is attenuated and distorted as it is propagated along the line and, to restore the pulse waveform, pulse regenerators are fitted at 1.828 km intervals along the length of the line.

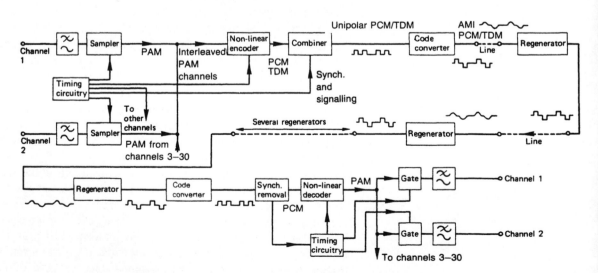

Fig. 12.26 30-channel PCM system

Higher-order PCM Systems

The ITU−T 30-channel PCM system is used as the basic building block for higher-order TDM systems which comprise the digital bearer network. A 30-channel system contains $32 \times 8 = 256$ bits and so the line bit rate is

$$256 \times 8000 = 2048 \text{ kbits/s} \equiv 2 \text{ Mbits/s}$$

This system is used to connect a digital local telephone exchange to a digital trunk exchange.

The higher-order bit rates recommended by the ITU−T are

A 8448 kbits/s   ≡   8 Mbits/s
B 34 368 kbits/s ≡   34 Mbits/s
C 139 264 kbits/s ≡ 140 Mbits/s
D 564 992 kbits/s ≡ 565 Mbits/s
E 2.4 Gbits/s

The 30-channel group can be represented by the block diagram of Fig. 12.27(*a*). The box marked *muldex* represents the multiplexing and demultiplexing equipment required for both directions of transmission.

Four 30-channel groups can be combined to form a 120-channel 8448 kbits/s system (Fig. 12.27(*b*)). Each of the four groups is fed into a store, the contents of which are fed into line in sequence. It should be noted that 4×2048 is 8192 kbits/s. Some of the surplus bits are used for synchronization purposes and supervisory signals and the remainder are redundant.

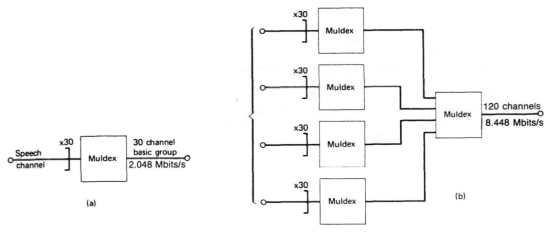

Fig. 12.27 Assembly of higher-order PCM systems: (*a*) 30-channel; (*b*) 120-channel

The inputs and outputs on the low bit rate side of an 8 Mbits/s or higher muldex are known as *tributaries*. The terminology is illustrated by Fig. 12.28. Signals passing over the interface between a 2 Mbits/s muldex and an 8 Mbits/s muldex are encoded into HDB3 code as are signals between 8 Mbits/s and 34 Mbits/s muldexes. The interfaces

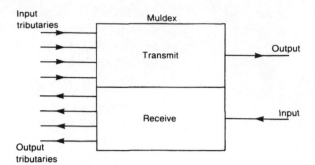

Fig. 12.28 Muldex terminology

used for 140 Mbits/s and higher muldexes employ a code known as *coded mark inversion* (CMI). This is a two-level non-return to zero code in which 0s are indicated by a low level followed by a high level, each lasting for a half-bit period, and 1s are indicated by alternate high and low levels, each for a whole-bit period. An example of the use of CMI is given in Fig. 12.29.

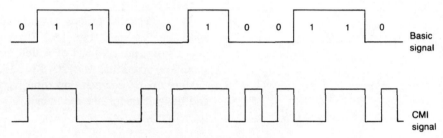

Fig. 12.29 Coded mark inversion

A muldex has two basic functions in each direction of transmission:

(a) Combining the bit streams arriving from four input tributaries that have the same nominal bit rate but, since their clocks are each subject to a tolerance, not necessarily identical rates. This is a process known as *justification* and it involves changing the bit rate of each tributary, by adding bits, so that each is at exactly the same bit rate.

(b) Insertion of a *frame alignment word* (FAW) whose function is to ensure that the composite signal arriving at the receiver is correctly demultiplexed.

(c) The demultiplexing of the input composite bit stream to its four output tributaries.

(d) To make use of the FAW to obtain continuous frame alignment.

In similar manner, four 8 Mbits/s systems can be combined to form a 34 Mbits/s system, four 34 Mbits/s systems can be multiplexed to give a 140 Mbits/s system, and so on, through stages of $4 \times 140 = 565$ Mbits/s, and $4 \times 565 = 2400$ Mbits/s. The way in which the higher-order systems are built up is shown by Fig. 12.30.

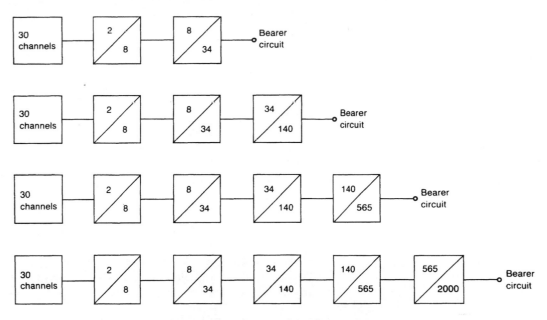

Fig. 12.30 Higher-order PCM systems

At each stage of multiplexing the frame structure of the composite bit stream has extra bits added to allow frame alignment to be maintained. One consequence of this is that it is not possible to identify a single 2 Mbits/s system and, if required, extract it from the higher-order system. If a 2 Mbits/s system is to be added to, or extracted from, a higher-order system it is necessary to demultiplex the whole system right down to the basic 30-channel system. This is shown by Fig. 12.31.

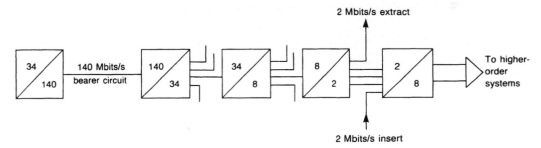

Fig. 12.31 Insertion and extraction of 2 Mbits/s PCM systems

Synchronous Digital Hierarchy

The *synchronous digital hierarchy* (SDH) is a worldwide standard for high-capacity digital systems that allows high-order multiplexed systems to be switched directly instead of having first to be demultiplexed down to the primary group stage. For direct switching to be possible the high-order muldexes must transmit in synchronism with the clocks in the digital telephone exchanges. Three stages of *synchronous transfer module* (STM) are used, as shown by Fig. 12.32.

Fig. 12.32 Synchronous digital hierarchy

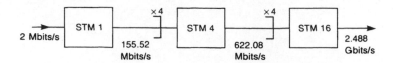

All the signals used in the SDH are synchronous and, in contrast to the plesiochronous DPH hierarchy (i.e. the PCM system hierarchy described earlier), SDH does not allow any fluctuations in the bit rate to occur. The muldexes and optical fibre terminals use the same frame structures as the digital exchange switches. The basic SDH signal is at 155.52 Mbits/s and it is known as STM1. Higher bit rates are integral multiples of 155.52 Mbits/s such as 622.08 and 2488.32 Mbits/s. Because it is a synchronous system SDH allows 2 Mbits/s groups to be added or extracted relatively easily using a *drop and insert equipment*, see Fig. 12.33. The cross-connect equipment allows the cross-connection of any 2 Mbits/s, or even 64 kbits/s, blocks of data between two high-order systems without any need for demultiplexing. The line coding employed is CMI.

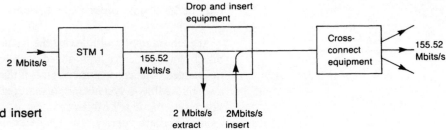

Fig. 12.33 Use of drop and insert equipment

Data Signals over PCM Systems

Each PCM channel operates at 64 kbits/s and the 30-channel PCM system operates at 2.048 Mbits/s and hence a number of data circuits can be accommodated. An individual channel can be used to provide a 64 kbits/s data circuit that can use a multiplexer to allow this circuit to transmit several lower-speed data circuits. Alternatively, a 2.048 Mbits/s PCM system can be used to provide a high-capacity data circuit that again will probably use multiplexers to allow the circuit to be used by a number of data circuits. The transmission of data circuits over PCM systems is considered in Chapter 14 (p. 210).

Relative Merits of FDM and TDM

The advantages of TDM, assuming the use of PCM, over FDM are as follows.

(*a*) The use of regenerators allows an almost distortion-free path regardless of distance, whereas in an FDM system the signal-to-noise ratio gets progressively worse with increase in distance.

(*b*) Channel selection is achieved by the use of relatively cheap electronic gates and low-pass filters rather than with expensive crystal filters.

(c) The level and phase of the received signal do not depend upon the stability of gain and/or phase shift of the circuits it has been transmitted through.

(d) In an FDM system, care must be taken to avoid non-linearity in amplifiers, etc., because this will result in intermodulation and crosstalk between channels. This is not so for a TDM PCM system.

The main disadvantage of PCM TDM is that a greater bandwidth is needed to transmit a given number of channels than is required for the corresponding FDM system, although this is not a problem if optical fibre cable is used. If the signal-to-noise ratio on a PCM transmission path falls below 21 dB, the received signal will deteriorate rapidly and become worse than the corresponding FDM signal.

Exercises

12.1   Draw, and explain the operation of the block schematic diagram of a three-channel PCM system. Ensure that the common line shows signals from each of the three different channels being transmitted.

12.2   The signal shown in Fig. 12.34 is sampled at the times, *a, b, c, d* and *e*. Show the binary-coded representation of each sample.

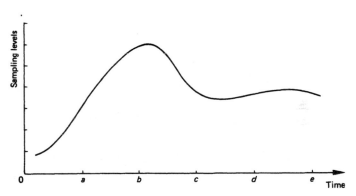

Fig. 12.34

12.3   The signals shown in Figs 12.35(*a*) and (*b*) are applied respectively to the channels of a two-channel PCM system. Sketch the waveform on the common line. Assume eight sampling levels are employed.

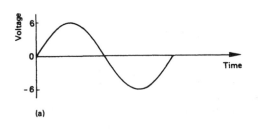

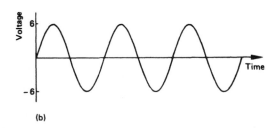

Fig. 12.35   (a)   (b)

12.4   Explain why accurate synchronization of a TDM system is essential.

12.5   Draw the frame structure of a 30-channel PCM system. State which of the channels are available for the transmission of speech. What do the other channels carry? Draw the frame structure.

12.6   Explain why the first step in the production of a PCM signal is the production of a PAM waveform.

12.7   Explain how quantization noise is generated within a PCM system. How can this form of noise be reduced and what is the disadvantage of so doing?

12.8   Explain why the range of signal amplitudes that a PCM system can handle is divided into a number of sampling levels. Why is the number of levels provided always some multiple of 2? How many bits are required to represent (*a*) 256 levels, (*b*) 8192 levels?

12.9   Successive samplings of a signal produce quantized levels of 3, 37, 53, 113, 60 and 28. Draw the binary waveform that is transmitted.

12.10   A PCM system has 1024 sampling levels. Calculate the number of pulses transmitted per second if the sampling frequency is 15 kHz.

12.11   The 30-channel PCM system uses pulses that are $0.488 \, \mu s$ wide. If the sampling frequency is 8 kHz and there are 256 sampling levels, calculate how many channels can be transmitted. Why are only 30 of these channels employed for speech transmission?

12.12   Calculate the line bit rate of a 30-channel PCM system. What would it be if 8192 levels were used?

12.13   Refer to Fig. 12.36. (*a*) What is a muldex? (*b*) Quote the higher-order bit rates that are recommended by the ITU−T. (*c*) Show how Fig. 12.36 can be made the basis of each of the higher-order systems in (*b*).

Fig. 12.36

12.14   List the relative merits of FDM and TDM.

12.15   A 1500 Hz sine wave of peak value 10 V is transmitted over a PCM system. The sampling frequency is three times the minimum allowable. Determine the instantaneous values of the sine wave at each sampling instant, assuming the first sample is taken at time $t = 0$. If the system caters for a maximum input voltage of ± 12 V and uses 128 sampling levels, write down the binary train transmitted to line to represent each sample.

12.16   Which quantum level represents a sampled amplitude of 0.54 V if each equally spaced quantum level corresponds to a voltage step of 0.01 V?

12.17   The number of quantum levels employed in a PCM system is increased from 128 to 512. Calculate the percentage reduction in the quantization error that results

12.18   The basic block diagram of a PCM system consists of the following blocks: (*a*) sampling gates; (*b*) encoder; (*c*) transmission line; (*d*) decoder; (*e*) demultiplexer; and (*f*) channel low-pass filters. Sketch the block diagram and indicate the waveform at the output of each block.

13 Wideband Radio Systems

Line-of-sight radio-relay systems form an integral part of the trunk network of most countries and their traffic capacity ranges from hundreds to several thousands of channels. Radio-relay systems operate in both the UHF and SHF frequency bands. Table 13.1 lists the systems that are in operation in the UK.

The BT microwave radio network has more than 200 relay stations that include both analogue and digital systems, although all new systems are digital. Analogue systems employ frequency modulation and digital systems employ 8-phase PSK, 16-state QAM, 64 QAM, or quaternary PSK (QPSK). A version of QPSK in which the QPSK signal is heavily filtered, known as *reduced bandwidth QPSK* (RBQPSK), is also employed. 64 QAM, for example, is able to

Table 13.1 Radio-relay frequency bands

Frequency (GHz)	Name	Use
3.7–4.2	4 GHz band	140 Mbits/s and
5.85–6.425	Lower 6 GHz band	2 × 34 Mbits/s digital
6.425–7.11	Upper 6 GHz band	140 Mbits/s, 565 Mbits/s
10.7–11.7	11 GHz band	and 2 × 34 Mbits/s digital
14.0–14.5	14 GHz band	Television.
17.7–19.7	19 GHz band	8 Mbits/s digital medium-capacity feeders. 140 Mbits/s and 565 Mbits/s digital

Notes:
A 140 Mbits/s system gives 1920 64 kbits/s channels.
A 565 Mbits/s system gives 7680 64 kbits/s channels.
The 2 × 34 Mbits/s systems give 720 64 kbits/s channels.

provide between six and eight bothway 140 Mbits/s channels. Six 140 Mbits/s channels gives $6 \times 1920 = 11\,520$ telephone channels. A more complex modulation system, known as 256 QAM, allows 565 Mbits/s channels to be transmitted. Because digital transmission is able to provide a satisfactory service (i.e. a small bit error rate), with a system signal-to-noise ratio of about 30 dB, as opposed to about 50 dB for an analogue system, it is often possible to transmit separate channels in the same frequency band using both horizontal and vertical polarization.

Radiation from an Aerial

Whenever a current flows in a conductor, the conductor is surrounded by a magnetic field, the direction of which is determined by the direction of current flow. If the current changes, the magnetic field will change also. Now, a varying magnetic field always produces an electric field that exists only while the magnetic field continues to change. When the magnetic field is constant the electric field disappears. The direction of the electric field depends on whether the magnetic field is growing or collapsing and can be determined by the application of Lenz's law. Similarly, a changing electric field always produces a magnetic field; this means that a conductor carrying an alternating current is surrounded by continually changing magnetic and electric fields that are completely dependent on one another. Although a stationary electric field can exist without the presence of a magnetic field and vice versa, it is impossible for either field to exist separately when changing.

If a sinusoidal current is flowing in a conductor the electric and magnetic fields around the conductor will also attempt to vary sinusoidally. When the current reverses direction the magnetic field must first collapse into the conductor and then build up in the opposite direction. A finite time is required for a magnetic field and its associated electric field to collapse, however, and at frequencies above about 15 kHz not all the energy contained in the field has returned to the conductor before the current has started to increase in the opposite direction and create new electric and magnetic fields. The energy left outside the conductor cannot then return to it and instead, is propagated away from the conductor at the velocity of light (approximately 3×10^8 m/s), see Fig. 13.1. The amount of energy radiated from the conductor increases with increase in frequency, since more energy is then unable to return to the conductor.

The energy radiated from the conductor or aerial, known as the *radiated field*, is in the form of an *electromagnetic wave* in which there is a continual interchange of energy between the electric and magnetic fields. In an electromagnetic wave the electric and magnetic fields are at right angles to each other and they are mutually at right angles to the direction of propagation, as shown in Fig. 13.2 for a particular instant in time. The plane containing the electric field and

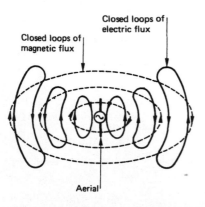

Fig. 13.1 Radiation from an aerial

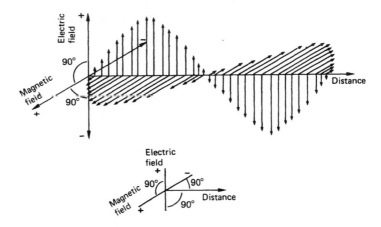

Fig. 13.2 The electromagnetic wave

the direction of propagation of the electromagnetic wave is known as the *plane of polarization* of the wave. For example, if the electric field is in the vertical plane, the magnetic field will be in the horizontal plane, and the wave is said to be vertically polarized. A vertically polarized wave will induce an e.m.f. in any *vertical* conductor that it passes, because its magnetic field will cut the conductor, but will have no effect on any horizontal conductor.

In the immediate vicinity of an aerial the electric and magnetic fields are of greater magnitude and different relative phase than in the radiated field. This is because there is, in addition to the radiated field, an *induction field* near the aerial. The induction field represents energy that is not radiated away from the aerial, i.e. the energy that does succeed in returning to the conductor, and its magnitude diminishes inversely as the square of the distance from the aerial. The magnitude of the radiated field is proportional to the frequency of the wave and inversely proportional to the distance from the aerial. Near the aerial, the induction field is larger than the radiation field, but the radiation field is the larger at distances greater than $\lambda/2\pi$, where λ is the wavelength of the signal radiated from the aerial.

The amplitudes of the electric field E and the magnetic field H in an electromagnetic wave bear a constant relationship to each other. This relationship is known as the *impedance of free space* and is the ratio of the electric field strength to the magnetic field strength, i.e.

$$\text{Impedance of free space} = \frac{E \text{ (volts/metre)}}{H \text{ (ampere-turns/metre)}}$$

$$= 120\pi \text{ ohms} \tag{13.1}$$

$$= 377\,\Omega \tag{13.2}$$

It is customary to refer to the amplitude of a radio wave in terms of its electric field strength.

Example 13.1

The magnetic field strength produced at a distance of 15 km from a transmitting aerial is 0.08 mA/m. Calculate (*a*) the electric field strength at the same point and (*b*) the electric field strength 30 km from the aerial.

Solution
(*a*) $E_{15} = 377 \times 0.08 \times 10^{-3} = 30.2$ mV/m   (*Ans.*)
(*b*) $E_{30} = 30.2/2 = 15.1$ mV/m   (*Ans.*)

Radio-relay Systems

Radio-relay systems using line-of-sight transmissions in the UHF and SHF bands can provide a large number of telephone channels and/or a television signal. Radio-relay systems are operated in the upper part of the UHF band and in the SHF band because it is then possible to provide a bandwidth of several megahertz. A wideband system is needed to accommodate several hundreds of telephone channels and/or a television channel.

A line-of-sight radio path uses a mode of propagation known as the *space wave*. The space wave travels in a very nearly straight-line path from the transmitting aerial to the receiving aerial as shown by Fig. 13.3. Nearly always the length of a route is longer than the line-of-sight distance over which the space wave is able to travel. It is then necessary to employ a number of radio-relay stations at intervals along the route. The function of a relay station is to receive the incoming radio signal and re-transmit it at a higher power level. The basic concept of a radio-relay system is shown by Fig. 13.4. The radio signal transmitted by the transmitting station is received by the first relay station where it is amplified (and pulse regenerated in a digital system), before it is re-transmitted to the next relay station. At this station the signal is again amplified (and regenerated) and re-transmitted and so on until the signal arrives at the receiving station.

Many of the radio-relay systems in use employ analogue techniques and hence the signal-to-noise ratio gets progressively worse with increase in the length of the system. The more modern radio-relay systems employ digital techniques with pulse regeneration at each relay station. This means that a required signal-to-noise ratio can be maintained over the length of the system.

At the transmitting station, the baseband signal (the signal produced by a PCM multi-channel system or by a television camera) is pre-emphasized and is then used to frequency-modulate a 70 MHz carrier. The modulated wave is then shifted to the allocated part of the

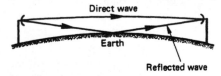

Fig. 13.3 The space wave

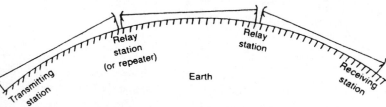

Fig. 13.4 Microwave radio relay system

frequency spectrum and amplified before it is radiated by the parabolic dish aerial. At each relay station, the received signal has its frequency changed back to 70 MHz before it is amplified, and perhaps regenerated, and it is then shifted to the frequency band to be used over the next section of the route. At the receiving station, the signal is shifted back to the 70 MHz band before it is demodulated.

Figures 13.5(*a*) and (*b*) show the basic block diagrams of a digital radio-relay system's transmitter and receiver, respectively. The digital signal applied to the transmitter is applied to the baseband processor. Here a clock frequency is established and the line coding (HDB3, CMI or AMI) is removed. The digital signal is then scrambled by a scrambler to break up any repetitive bit patterns that might exist and a frame is set up. The scrambled signal is then split into two 140 Mbits/s bit streams for QPSK, four bit streams for 16 QAM, and so on, and each of these bit streams is applied to the modulator. The output of the modulator is amplified and filtered before it is combined with the outputs of other channels and passed on to the r.f. amplifier. The combined signal is amplified and then passed on to the transmitting aerial to be radiated towards the destination station.

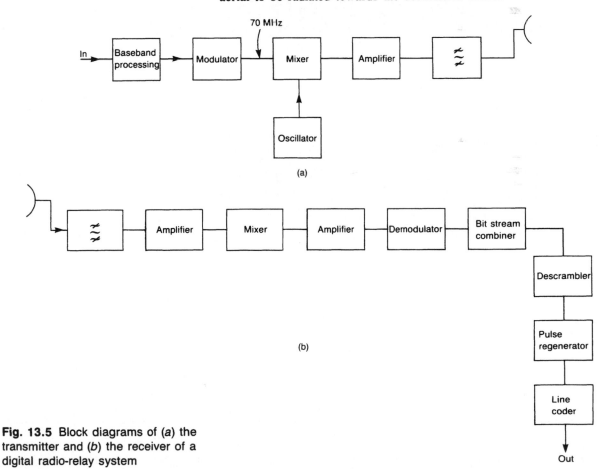

Fig. 13.5 Block diagrams of (*a*) the transmitter and (*b*) the receiver of a digital radio-relay system

The signal arriving at the receiver is selected by the channel filter and applied to a mixer. Here it is mixed with a 1 GHz carrier wave and the lower 140 MHz sideband is selected and applied to the demodulator. The demodulated output consists of two, or more, 140 Mbits/s bit streams and these must first be recombined into a single bit stream, and then descrambled to recover the original digital signal. This is then applied to a pulse regenerator before the appropriate line coding is carried out.

The modulation methods used for the different radio-relay systems are listed in Table 13.2. Figure 13.6 shows how 2×34 and 140 Mbits/s PCM systems are transmitted over 4 and 11 GHz systems respectively.

Table 13.2

Channel capacity (Mbits/s)	Frequency band (GHz)	Modulation method
2×34	4	8 PSK
2×34	lower 6	8 PSK
2×34	upper 6	8 PSK
2×34	11	8 PSK
2×34	14	8 PSK
140	4	16 QAM
140	4	RBQPSK
140	lower 6	RBQPSK
140	upper 6	16 QAM
140	11	QPSK/8 PSK/16 QAM
140	14	QPSK
8	19	QPSK
140	19	QPSK

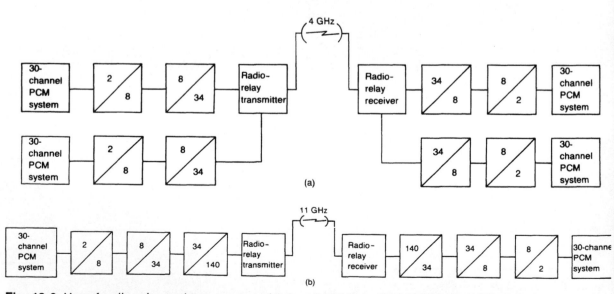

Fig. 13.6 Use of radio relay systems to carry: (*a*) 2×34 Mbits/s, (*b*) 140 Mbits/s PCM systems

Both radio relay and PCM line systems are widely used as integral parts of the national telephone network. The two systems have a number of advantages and disadvantages relative to one another which often means that one or the other is best suited for providing communication over a given route. The relative merits of the two systems are listed below.

(a) A radio-relay system is generally quicker and easier to provide.
(b) The problems posed by difficult terrain are easier to overcome using a radio-relay system.
(c) It is easier to extend the channel capacity of a radio-relay system.
(d) Difficulties may be experienced in obtaining suitable (line-of-sight distance) sites for a radio-relay system.
(e) When relay station sites have been chosen it may be difficult to gain access to them, whereas coaxial systems usually follow roads and so access is relatively easy.
(f) The transmission performance of a radio-relay system is adversely affected by bad weather conditions.

Communication Satellite Systems

Most of the long-distance international telephone traffic that is not carried by submarine cable systems is routed via a broadband communication satellite system, the basic principle of which is illustrated by Fig. 13.7. The ground stations are fully integrated with their national telephone networks and, in addition, the European ground stations are fully interconnected with one another. Four frequencies are used; the North American ground station transmits on frequency f_1 and receives frequency f_4, the European stations transmit frequency f_3 and receive frequency f_2. Essentially, the purpose of the communication satellite is to receive the signals transmitted to it, frequency translate them to a different frequency band (f_1 to f_2 or f_3 to f_4), amplify the signals, and then re-transmit them to the ground station at the other end of the link.

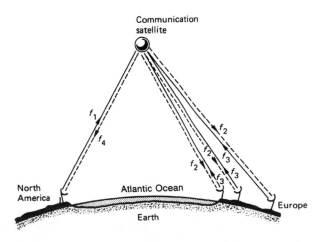

Fig. 13.7 Communication satellite system

Communication satellites which form an integral part of the public international telephone network are operated on a global basis by COMSAT (Communication Satellite Corporation) on behalf of an international body known as INTELSAT (International Telecommunication Satellite Consortium). The COMSAT system employs communication satellites travelling in the circular equatorial orbit at a height of 35 880 km. This particular orbit is known as the synchronous orbit because a satellite travelling in it appears to be stationary above a particular part of the Earth's surface. Seven satellites are used, positioned around the Earth so that nearly all parts of the surface of the Earth are 'visible' from at least one satellite. A large number of ground stations are in use and now number more than a hundred in nearly a hundred different countries.

Each ground station transmits its telephone traffic to a satellite on the particular carrier frequency allocated to it in the frequency band 5.925—6.425 GHz or 14.0—14.5 GHz. This is a bandwidth of 500 MHz and allows the simultaneous use of a satellite by more than one ground station. Different ground stations are allocated different carrier frequencies within this 500 MHz band, either permanently or for particular periods of time — depending on the traffic originated by that station. This method of utilizing the capacity of a communication satellite is known as *frequency division multiple access* (FDMA). Each of the allocated carrier frequencies has a sufficiently wide bandwidth to allow a multi-channel telephony system to be transmitted and, in some cases, a television channel. The number of telephony channels thus provided varies from 24 in a 2.5 MHz bandwidth to 1872 in a bandwidth of 36 MHz. All the signals transmitted by a satellite are transmitted towards every ground station and each station selects the particular carrier frequencies allocated to it in the band 3.700—4.200 GHz or 10.95—11.2 GHz and 11.45—11.7 GHz.

Exercises

13.1 State the mode of propagation that is used at UHF and SHF for terrestrial radio-relay systems. What is the approximate maximum distance of a radio link using this mode? How is the distance extended? State the frequencies employed and explain why different frequencies are used for each direction of transmission.

13.2 Draw the block diagram of a digital radio-relay transmitter and explain its operation. Outline the QAM 16 modulation method.

13.3 Draw the block diagram of a digital radio-relay receiver and explain its operation. Why is pulse regeneration employed?

13.4 Describe, with the aid of diagrams, how PCM systems are transmitted over a radio-relay system. Which bit rates are employed? How many telephone channels can each bit rate transmit?

13.5 Draw the basic set-up of a satellite communication system and explain its operation. Why are four different frequencies used for each satellite link?

14 Data Circuits

Many firms and organizations have a number of factories, warehouses, offices and other establishments where members of staff are employed. Usually, many of these locations are not sited at the head office, indeed some may be sited many hundreds or even thousands of kilometres away. Any of the remote locations may need to send information to the head office at regular intervals varying from monthly or weekly to daily, or even more frequently. The information could, of course, be sent by post but there is an ever increasing demand by management for information to be made available more quickly. The answer to the problem can sometimes be given by the Telex system, but this is limited in its maximum speed of transmission, and some faster system is often required. Much of the data required by a head office is destined to be stored and/or processed by a digital computer, and computers are employed for a great number of varied purposes in both the industrial and the commercial fields. Computers are used to carry out complex calculations, to maintain and update records such as technical data, tax records and medical histories. Computers are also used in the preparation of weather reports and forecasts, the calculation and printing of wages and salaries, and preparation of bills and invoices; they are also used for inventory control, and for the control of industrial processes in factories. The high-street banks and building societies make use of computers to maintain details of customers' accounts, of standing orders, direct debits, etc., and to operate cash machines, while airlines and package holiday firms are able to operate booking systems that provide an immediate confirmation of vacancies and bookings.

The many applications of the digital computer have led to most organizations investing in their own computer facilities. A mainframe digital computer is an expensive purchase; it is not economically possible for a large organization to install such a computer at several locations, and small firms cannot afford one at all. The vast majority of firms and companies, big and small, will employ one micro-computer at least, and very often several. Most microcomputers are either manufactured by IBM or they are IBM-compatible and these computers are known as *personal computers* (PCs). Apple MacIntosh microcomputers are not IBM-compatible but it is fairly common practice to refer to all microcomputers as PCs.

The facilities provided by a mainframe computer may be needed at many different points in the organization's set-up. The computing facilities needed at different points are sometimes provided by a microcomputer but most often by a number of PCs. There is often a requirement for PCs to communicate with one another or with the mainframe computer. Because of this there is a considerable demand for data links, to connect data terminals with the mainframe host computer and so extend the use of the mainframe computer to a number of distant locations. The term 'data terminal' is used to include both PCs and visual display units (VDUs) which consist essentially of a monitor and a keyboard.

The Use of Leased Circuits and the PSTN

A data link that connects a data terminal to a remote digital computer may be leased from the telephone administration or it may be temporarily set up by dialling a connection via the PSTN. The data link is said to be analogue because the data terminal is connected to the digital core network by a local line in the analogue access network. A modem is necessary at each end of the circuit.

The choice between leasing a private circuit and using the PSTN must be made by the careful consideration of factors such as the cost, availability, speed of working, and transmission performance. Private circuits may transmit digital signals over Kilostream or Megastream circuits (see p. 211) or may use modems to convert the data signals into voice-frequency signals before transmission over the access network.

Voice-frequency data circuits may also be of any length and are routed over multi-channel PCM systems. (A multi-channel PCM system may itself be routed, wholly or partly, over a microwave radio link.)

A point-to-point privately leased circuit provides exclusive use of the transmission path and so it is instantly available. A PSTN connection is established by dialling the telephone number of the distant data terminal and the route used for a particular call is a random choice from a large number of different possible routings. There will always be some delay before a PSTN connection is set up and such a delay may not be acceptable.

Now that the core network is all-digital any degradation of the transmission performance of an analogue data circuit is most likely caused by a local line in the access network. Transmission at bit rates of up to 14 400 bits/s or more is available over the PSTN but the results cannot be guaranteed; some of the dialled connections may not be adequate and the error rate may be too high. High-speed modems have one, or more, lower fall-back speeds that may be used when this occurs. One modem standard offers a bit rate of 14 400 bits/s, with reduced speed facilities at both 12 000 bits/s and 9600 bits/s.

Half-duplex operation only is usually offered over the PSTN since full-duplex operation normally requires the use of a four-wire presented circuit. If full-duplex working is wanted then a dedicated circuit must be provided or a modem that incorporates echo cancellation must be employed.

The bandwidth provided by a leased circuit is 300—3000 Hz, the lines being adjusted to give a good transmission performance over this bandwidth. At higher frequencies group delay—frequency distortion increases rapidly. The use of a privately leased, or *dedicated*, circuit for data transmission has the following advantages over the use of the PSTN.

(*a*) Exclusive use of the circuit is obtained. Thus time is not wasted setting up calls and the performance of the circuit remains stable.

(*b*) The link can be adjusted to have the optimum performance and it is less affected by noise and interference. As a result higher speed, more reliable, transmissions may be possible.

(*c*) Full-duplex operation is available if the dedicated circuit is four-wire presented, and higher data 'throughput' is possible by eliminating turn-round time. (With half-duplex operation the time taken for the modem to switch from reception to transmission and vice versa — known as the *turn-around time* — is some tens of milliseconds.)

On the other hand, the cost of permanently leasing a line from the telephone company may be relatively high and it may only be economically justified if there is sufficient data traffic on the line or if the particular terminal application necessitates a permanent connection, e.g. a cash dispensing terminal in a bank which checks a customer's account before releasing the cash. If the data communication requirements involve occasional contact with a large number of locations, and the majority of the connections are of fairly short duration, the use of the PSTN is probably best. On the other hand, if long duration connections between a few data terminals and the host computer centre are planned for, leased links will probably be chosen. In practice, most private data networks consist of a combination of both leased and PSTN links, and very often the leased circuits are provided with the stand-by facility of using the PSTN when necessary (i.e. should the dedicated circuit fail).

Two-wire and Four-wire Presented Circuits

The terms two-wire and four-wire refer to the line circuit that is presented to the data terminal.

Two-wire Presented Circuits

Figure 14.1 shows a two-wire presented circuit. The trunk circuit set up via the core network is routed via one, or more, PCM systems. Different channels are employed for each direction of travel so that the trunk circuit is four-wire operated. But since the local lines

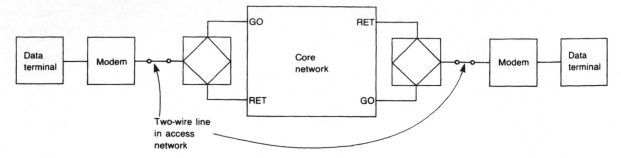

Fig. 14.1 Two-wire presented data circuit

connecting the data terminals to the core network are operated two-wire, the circuit as a whole is said to be two-wire presented.

When FSK is employed the two-wire presented ends of the circuit are only able to transmit 600 bits/s or 1200 bits/s signals in one direction at a time (half-duplex) since the frequencies used are 1300 Hz and either 1700 Hz or 2100 Hz and these occupy most of the available frequency band. Full-duplex operation is only possible if the return direction of transmission is operated at a different frequency and at a lower bit rate. Alternatively, full-duplex operation is possible if both directions of transmission are operated at the lower speed of 300 bits/s. The two directions of transmission are operated in different frequency bands and so they can be sorted out at the receiver by means of suitable filters.

These low bit rate FSK systems are rarely, if ever, employed in modern data communication networks.

Four-wire Presented Circuits

When a data link is operated on a four-wire presented basis, two pairs of conductors are extended from the core network right up to the modem in the data terminal as shown by Fig. 14.2. The data circuit now provides two separate channels between the two data terminals and each channel is able to carry high-speed data at the same time. Thus, the four-wire presented circuit provides full duplex high-speed operation.

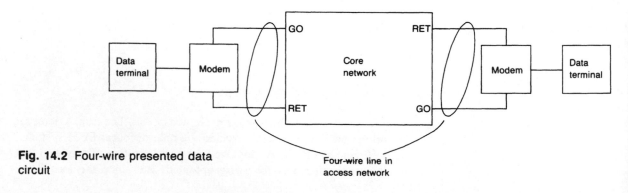

Fig. 14.2 Four-wire presented data circuit

Multiplexers, Front-end Processors and Concentrators

Because of the high costs of long-distance trunk circuits, whether leased or dial-up, it is desirable to be able to use a telephone circuit to carry more than one data link. Multiplexers and concentrators, multi-drops and cluster controllers, are equipments which, in their different ways, can increase the utilization of a point-to-point circuit. Further savings in costs are obtained with a concentrator since fewer modems and interfacing equipments are then required. A front-end processor relieves a mainframe computer of many tasks and allows time-sharing of the computer.

Multiplexers

A *multiplexer* combines several individual data links together using either *frequency-division multiplex* (FDM) or *time-division multiplex* (TDM). FDM techniques are only used for *data-over-voice* circuits and broadband *local area networks* (LANs).

With a time-division multiplexer, a time slot on a high-speed bearer circuit is allotted to each channel in time sequence. Figure 14.3 illustrates the basic principle of a time-division multiplexer. Four non-synchronous data channels are shown operating at 2400 bits/s. The duration of a bit is 1/2400 s or 417 μs and so a 10-bit character occupies a time slot of 4.17 ms. If the bearer circuit is to be operated at the higher speed of 9600 bits/s each bit sent to line will have a time duration of 104.2 μs. The data present on the channel 1, 2, 3 and 4 inputs to the TDM system is fed into the appropriate channel buffer stores and is held there until the store is given access to the bearer circuit. The clock pulses 1, 2, 3 and 4 are applied to each gate in

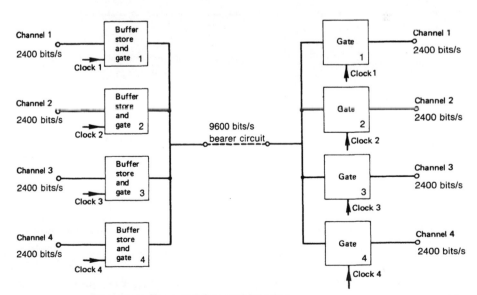

Fig. 14.3 Time-division multiplexing

turn to sample the stored information held in the associated buffer store and apply it to the bearer circuit for a time period equal to the duration of a character, i.e. 1.042 ms. When, for example, clock pulse 1 enables store 1 for 1.042 ms, the stored data is transmitted to line at the 9600 bits/s bit rate. Clock pulse 1 then ends and buffer store 1 is taken off line. Clock pulse 2 now enables store 2 to allow one character of the data stored in it to be sent to line. After 1.042 ms buffer store 2 is inhibited and buffer store 3 enabled, and so on until all four channels have been connected, in turn, to the bearer circuit. Clock pulse 1 now enables buffer store 1 again and another character of the stored data can be transmitted to the line and so on.

The start—stop bits are often *stripped* before the data is transmitted over the multiplexed circuit. Bit stripping increases still further the traffic utilization of the bearer circuit. The stripped start and stop bits must be replaced after the composite signal has been demultiplexed at the receiving end of the high-speed line. Clearly, synchronization between the transmitting and receiving equipments is essential in order that the clock pulses at the receiver occur at the correct intervals in time. Otherwise, the bits proper to one channel may well be routed to another channel.

Often, the output of a multiplexer will be applied to a modem to convert the multiplexed digital signals into voice-frequency signals for transmission over the telephone network, as shown by Fig. 14.4. With a TDM system the bit rate on the bearer circuit is the *sum* of the individual channel bit rates.

Example 14.1

A 14 400 bits/s data link is to carry one 4800 bits/s and a number of 2400 bits/s data channels using a multiplexer. Determine the number of 2400 bits/s channels that can be transmitted.

Solution
Capacity available for 2400 bits/s channels = 14 400 − 4800 = 9600 bits/s
Hence, number of 2400 bits/s channels = 9600/2400 = 4.   (*Ans.*)

Often the multiplexed data channels do not all operate at the same bit rate. When this is the case, (*a*) extra time slots can be allotted to the higher-speed channels, or (*b*) all the channels can be allotted the same time slots and then the lower-speed channels will not occupy all of their allocation.

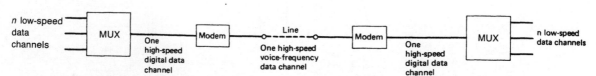

Fig. 14.4 Multiplexed voice-frequency data circuit (MUX = multiplexer)

Bit Interleaving

When synchronous data channels are to be multiplexed it is usual to employ *bit interleaving*. The multiplexed bit streams include all the synchronizing and control characters that are present with the data. Each incoming bit is transmitted straight onto the bearer circuit as it is received and hence the multiplexing terminals do not need large-capacity stores and the control circuitry is simpler.

Statistical Multiplexers

The number of data channels that can be transmitted over a bearer circuit may be further increased if a *statistical multiplexer* is employed. The operation of a statistical multiplexer is based upon the probability that at any instant in time only some of the terminals connected to the multiplexer will be active. Those terminals that are not transmitting or receiving data at that moment do not need to be allotted a time slot. The data carrying capacity of the bearer circuit is allotted dynamically to the active inputs and this practice typically allows the bearer circuit to have a bit rate of between one-half and one-quarter of the aggregate bit rate. An example of the use of a statistical multiplexer is shown by Fig. 14.5.

Front-end Processors

A mainframe computer generally interfaces with a data communication system through a *front-end processor* (FEP). The FEP is a smaller computer with software that allows it to relieve the mainframe computer of some tasks it would otherwise have to perform. This enables the mainframe computer's power to be devoted to the processing and storage of data. An FEP performs the following tasks.

(a) The FEP acts as a multiplexer to allow several data channels to have access to the computer on a time-sharing basis. One or more high-speed channels connect the FEP to the computer, while the other side of the FEP is connected to a greater number of data links.

(b) The FEP acts as a communications controller. It controls all the telecommunications facilities of the computer, monitoring all the

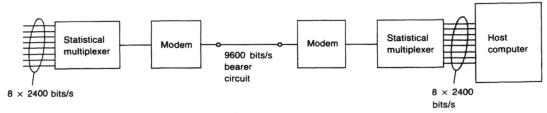

Fig. 14.5 Statistical multiplexer

associated modems to determine when any of them has data ready for processing. The FEP decides when a modem shall have access to the computer itself and so ensures that the computer is not overloaded by a large number of messages arriving at the same time. The FEP sets up all the necessary connections via the PSTN and meters all incoming calls and then, if appropriate, produces bills for the use made of the computing facilities.

(c) The FEP *polls* each data terminal in turn to determine whether it has any data to transmit to the computer. The interrogated terminal can reply by sending its data or it can signal that it has none to send. The FEP can then check whether it has any data to send to that data terminal and if so transmits it. The FEP will then poll the next terminal in the laid-down sequence. The polling can be carried out at such a speed that the FEP acts as a multiplexer to provide time-sharing of the computer.

Concentrators

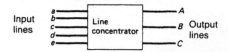

Fig. 14.6 Use of a line concentrator

The operation of a *concentrator* relies upon the fact that data are not normally transmitted continuously over a data link but, instead, are sent in *bursts* of varying lengths. A line concentrator is able to connect any of x inputs to any of y outputs, where $x > y$. The concept of contention is illustrated by Fig. 14.6 in which any one of the input lines a, b, c, d or e can be connected to any one of the output lines A, B or C. The concentrator polls each of its inputs in turn to discover whether it has data waiting to be transmitted. If it has, that input is switched through to an unoccupied output line and the distant modem is signalled that data transmission is about to commence. When incoming data are received via one of the lines A, B or C, the concentrator determines the destination data terminal and addresses the data.

Data concentrators include storage facilities which allow the input data to be stored and then read out from the store and transmitted down the line in blocks when the concentrator is polled by the FEP at the computer centre. The storage facility allows data to be sent in smooth blocks of data (which may well be the combination of the data originating from more than one terminal) and allows the use of a main line whose bit rate is less than the sum of the input bit rates. This does mean, of course, that if all the input lines are sending data at the same time, some of the data will have to be stored and there will be some delay in the concentrator before all of the data is passed on to the main line. Figure 14.7 shows how a data concentrator is used; it should be noted that the concentrator works with digital signals only and hence any analogue signals must be changed into digital form before entering the circuit.

A line concentrator may be used before a multiplexer to achieve the maximum utilization of the transmission path and Fig. 14.8 gives

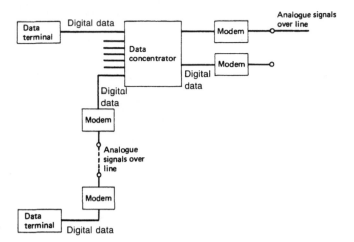

Fig. 14.7 Use of a data concentrator

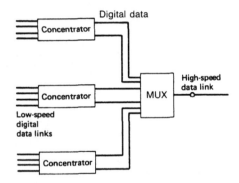

Fig. 14.8 Use of concentrators and multiplexers

an example of this technique. For each of the line concentrators the bit rate at its output is less than the total bit rate at its input, but for the multiplexer the output bit rate is higher than the input bit rate on each input.

A multiplexer can also be used as a simple kind of front-end processor to permit time-sharing of a digital computer and Fig. 14.9 shows a possible data network. The FEP polls each of the concentrators sequentially and during the short time a concentrator is connected to the computer it can pass data into, or receive data from, the computer. Clearly, the concentrator must possess some storage facility so that it can store the data received when the computer is dealing with another concentrator, or with a direct link. Some concentration may be provided at the location of the host computer

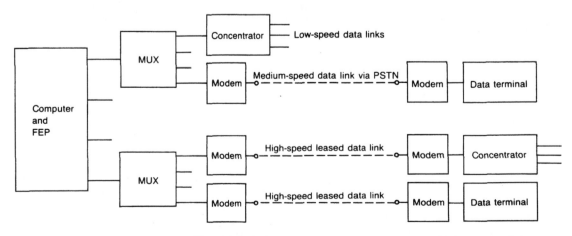

Fig. 14.9 A data network including both multiplexers and data concentrators

concentrators may be located at sites which are considerable distances away.

Multi-drop and Cluster Networks

Two other means of increasing the traffic-carrying capacity of a bearer circuit are the use of a *multi-drop circuit* and the use of a *cluster controller*. A multi-drop circuit allows several terminals to be connected to the data circuit at different points along its length as shown by Fig. 14.10. Clearly, the method is limited to lines that are not routed over a PCM system. A cluster controller is able to split a line into two, or more, branches as shown by Fig. 14.11. All the clustered terminals must be located within a few kilometres of the cluster controller.

Error Control

A data link is subjected to a number of sources of noise that can produce errors in the received data. Most serious are the errors

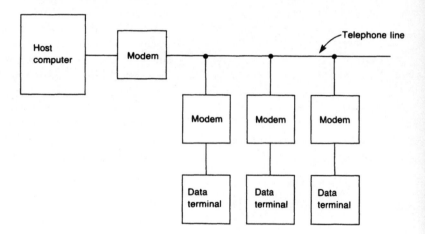

Fig. 14.10 Multi-drop data circuit

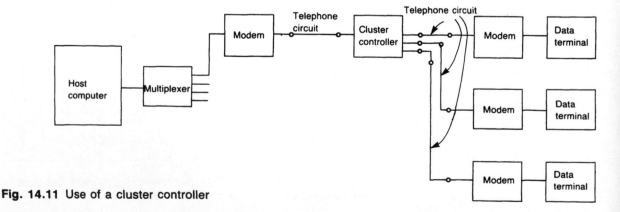

Fig. 14.11 Use of a cluster controller

produced by brief breaks in the transmission path; the loss of a single bit may well cause an error which could alter the whole meaning of a message. The possible causes of interruptions are many and include the following: poorly soldered joints, changes in the power supplies from the main plant to the standby plant, mechanical vibrations, and last, but by no means least, breaks caused by technicians working on nearby circuits or equipments.

To reduce the error rate to an acceptable level (it is not economically possible to eliminate it), a data system will usually include some kind of error-checking mechanism and in some cases error correction as well. A number of error detection methods are in use in different networks but this book will only outline the principles of the most commonly employed method, known as the *parity system*.

Character parity means that each character signalled to the line has one extra bit added. The added bit may be either a 1 or a 0; the choice being made so that the *total* number of 1 bits transmitted per character is *odd* if *odd parity* is used or *even* if *even parity* is employed. The parity bit is transmitted at the end of each character. Two examples of this system are given below.

(*a*) Character 0 0 1 1 1 0 1; there are four 1 bits in this character and so the added parity bit must be a 1 if odd parity is used and a 0 if even parity is employed.

(*b*) Character 1 0 1 0 1 1 1; there are five 1 bits in this case and so odd parity requires the addition of a 0 parity bit, while if even parity is used the added parity bit must be a 1.

At the receiver, the incoming character is checked to determine how many 1 bits it contains, and if the total is odd (or even) the received character is taken as being correct. Should the total number of 1 bits received be an even number when odd parity is used, or an odd number when even parity is used, an error in the received message is detected which must, in some way, be corrected at the receiving terminal or signalled back to the transmitter so that the data can be sent again.

Parity bit error checking is successful in locating any single-bit errors that should occur although clearly two bits in error would not be detected. However, the likelihood of two bits in error in one character is acceptably small.

Two-bit errors can also be detected if the parity bit principle is extended to whole blocks of data. It is also possible to use parity checks other than the number of 1 bits received per character but such procedures are beyond the scope of this book. In practice, block parity systems are normally used and the use of character parity is fairly rare.

With an error correction system it is usual to acknowledge each character or each block as it is received. Character-by-character acknowledgement is very time consuming when the error rate is low but, on the other hand, block error correction means repeating a complete block of data whenever an error does occur. For either error correction method, the next character of the next block is not sent by the transmitter until an ACK signal is received from the distant receiver.

Data Networks Using Modems

A data network will contain one or more digital computers and a number of data terminals, each of which will have some kind of access to a computer. The simplest data circuit consists of two modems connected together by a voice-frequency circuit routed over either the PSTN or a dedicated circuit. A point-to-point single-port circuit is shown by Fig. 14.12. If the data terminals are located at the same site as the modem, a modem-sharing unit can be used as shown by Fig. 14.13. If, however, the terminals are not at the same site but instead are located within a few kilometres of the modem's location, a cluster arrangement can be employed. A possible network is shown by Fig. 14.14, in which two dropped terminals are also shown. The computer will poll each terminal in sequence to check whether it has data to transmit to the terminal, or whether the terminal has data to send to it. Multi-drop and cluster networks provide a cheap and quite efficient method of providing data communication between different points when the remote sites are geographically sited within a few kilometres of one another.

The various methods of increasing the traffic utilization of a telephone circuit may be combined in various ways and Figs. 14.15(*a*), (*b*) and (*c*) shown three possible networks.

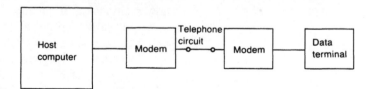

Fig. 14.12 Point-to-point data circuit

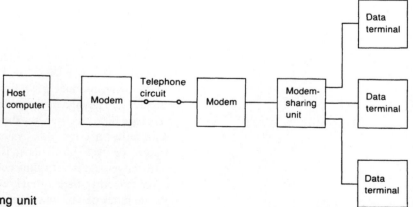

Fig. 14.13 Use of a modem-sharing unit

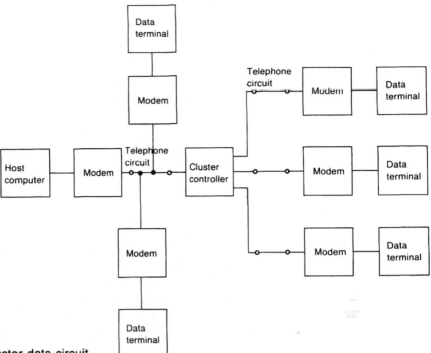

Fig. 14.14 Multi-drop and cluster data circuit

A more complex network is shown by Fig. 14.16. This network is a composite of the various techniques outlined earlier in this chapter.

One of the more common kinds of data network is the kind where data originating from a large number of variously located branch offices is to be processed by a central computer, stored, and when needed is transmitted back to the originating or some other branch office. This is, of course, the type of network used by the high-street banks. Each branch office has access to a central computer which keeps details of customers' accounts, standing orders, direct debits, and so on, and prints out bank statements, and pays standing orders and direct debits by transferring data from one location to another. When a payment is made from an account with one bank to an account held with another bank, a computer-to-computer transfer of data will be needed. When two mainframe computers are to communicate directly with one another they can be linked by a *Megastream* circuit, see Fig. 14.17.

Figure 14.18 shows the layout of a possible centralized accounting data system. The cluster controllers allow a number of data terminals to share a common line without any change in the bit rate. The terminals are polled by the FEP and are signalled when they can transmit their data. Should a branch be unable to establish contact with the computer over its direct route for some reason, a link can be set up instead over the PSTN.

Computers are also widely used by other organizations of many

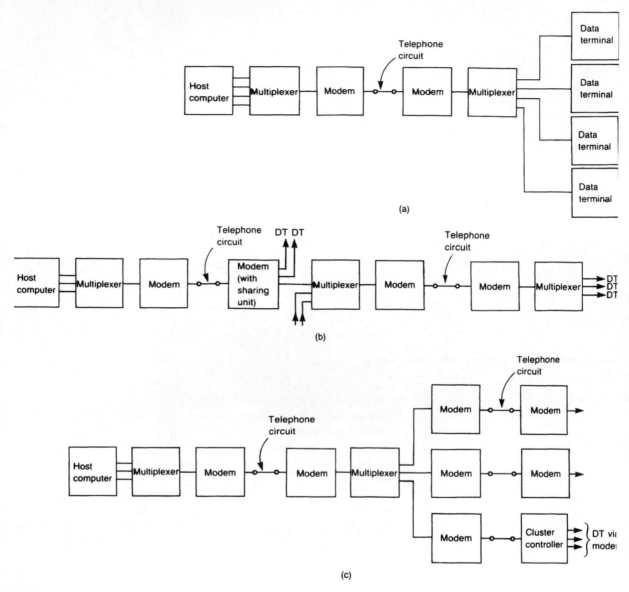

Fig. 14.15 Three possible data circuits using modems and multiplexers (DT = data terminal)

kinds, such as the gas, water, and electricity companies, and, o course, the police. In the UK, for example, a central police compute is used to store records of criminals and of stolen property such a motor cars, as well as details of missing persons. Clearly it i convenient for the police if every policeman, on foot, in a car or ii an office, can have immediate access to the computer to obtain any necessary information. The basic way in which this is achieved i shown by Fig. 14.19. All the police headquarters in the country have access to the centralized police computer and all policemen have acces

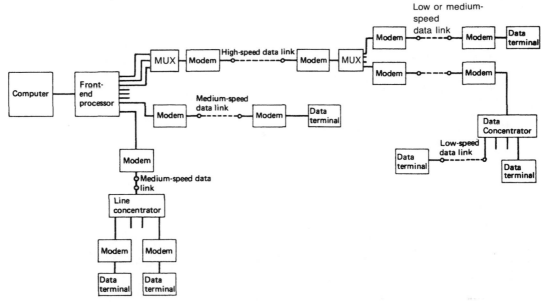

Fig. 14.16 On-line data circuit

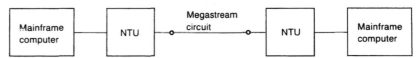

Fig. 14.17 Megastream circuit links two computers
(NTU = network terminating unit)

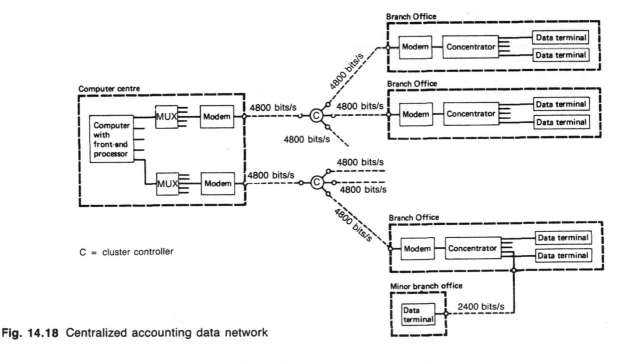

Fig. 14.18 Centralized accounting data network

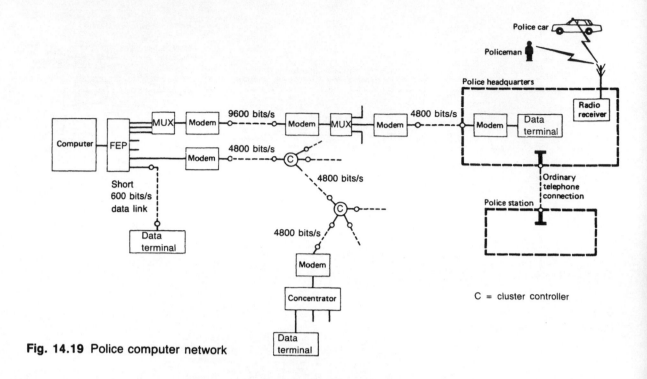

Fig. 14.19 Police computer network

to a data terminal. This access may be via a radio link or a telephone call as shown in the figure or, when it is convenient, by personal contact with the data operators.

Data Circuits over Digital Networks

The ITU−T 30-channel PCM system provides 30 64 kbits/s channels with the bearer circuit operating at 2.048 Mbits/s. These bit rates are much larger than the speeds the analogue network, especially the PSTN, can handle. This means that the standard PCM system has the capability to accommodate a number of high-speed data circuits. Each terminal of a speech or analogue PCM system includes circuitry for analogue-to-digital conversion (ADC) and for digital-to-analogue conversion (DAC). When, therefore, one or more of the 30 channels is to carry a data circuit the data signal must be converted to voice-frequency form before it is applied to the PCM channel input. This step is carried out so that the terminals need not be modified by having some of their ADC and DAC equipments removed. Thus, so far as the PCM system is concerned, it is only carrying a number of voice-frequency analogue signals in digital form. The arrangement employed is shown by Fig. 14.20(*a*). The 64 kbits/s channel provided can be multiplexed to provide an increased number of lower-speed channels as shown by Fig. 14.20(*b*).

If the entire 2.048 Mbits/s capacity of a PCM system is to be used for the transmission of digital signals the PCM terminal equipment that is employed is *not* fitted with either channel ADC or channel

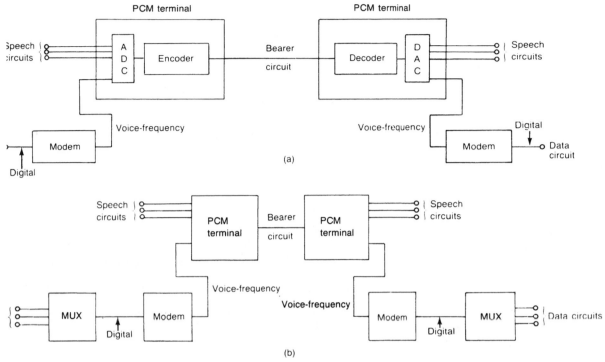

(a)

(b)

g. 14.20 Data signals routed over
ie channel in a PCM system

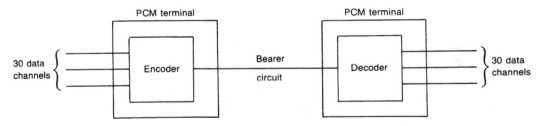

g. 14.21 Digital PCM system

DAC. The input digital signals are directly applied to the encoder and converted into the PCM format. At the receiving end of the system the PCM signals are directed to their appropriate channels, decoded, and then output in digital form.

Figure 14.21 shows the basic principle of the digital PCM system; each of the 64 kbits/s channels may be subdivided into a number of lower-speed channels by digital multiplexing equipment. This equipment can be provided at the site of the PCM terminal or at the premises of the user. BT offers digital services which are based upon the method shown by Fig. 14.21.

(a) *Kilostream* is a digital point-to-point service which is available at bit rates of 2400 bits/s, 4800 bits/s, 9600 bits/s, 48 kbits/s and 64 kbits/s. The service is provided using a 64 kbits/s channel and the appropriate multiplexing equipment (except at 64 kbits/s).

Kilostream offers a cheaper alternative to analogue transmission using modems.

BT's Kilostream (N) voice service provides a relatively cheap way of connecting a digital leased circuit to two private automatic branch exchanges (PABXs) to give voice facilities. Any number of 64 kbits/s channels from 2 to 16 can be selected to support a mix of speech and data applications. The service is intended to fill a gap between Kilostream and Megastream.

(b) *Megastream* offers a point-to-point digital service operating at either 2.048 Mbits/s, 8 Mbits/s, 34 Mbits/s or at 140 Mbits/s. The service can be used without multiplexing to provide a wide-band high-speed system or it can be multiplexed to give a number of lower-speed channels. The channels need not all carry data; if required some of the channels can carry digitized speech signals instead and Fig. 14.22 gives a typical arrangement.

(c) *Switchstream* offers full-duplex data communication at speeds of up to 48 kbits/s. It employs a technique that is known as packet switching.

(d) *Satstream* provides leased high-speed links via communication satellites to countries in Western Europe, to the USA and to Canada.

The use of a digital circuit to carry data signals offers a number of advantages over analogue transmission.

(a) Impulse noise is eliminated.
(b) White noise is not cumulative.
(c) Because of (a) and (b) data can be transmitted at higher bit rates.
(d) An expensive modem is not needed at each end of the data link.

Examples of the use of Kilostream and Megastream digital circuits are shown in Figs. 14.23 and 14.24. It can be seen that when Megastream circuits are employed telephony services via digital PABXs are often provided in addition to data communication.

The digital data circuits, Kilostream and Megastream, form the mainstay of the data networks installed in the UK. Often neither a

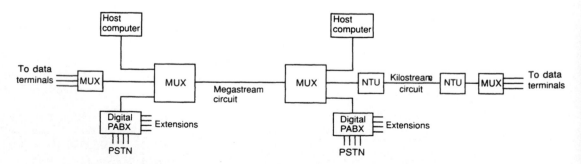

Fig. 14.22 Use of Megastream and Kilostream

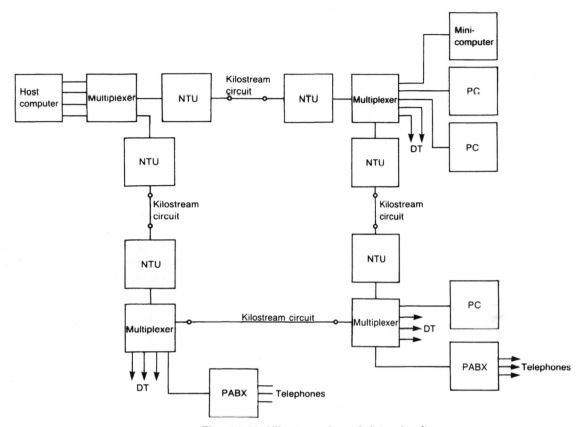

Fig. 14.23 Kilostream-based data circuit

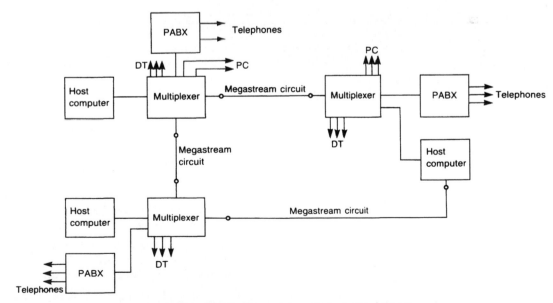

Fig. 14.24 Megastream-based data circuit

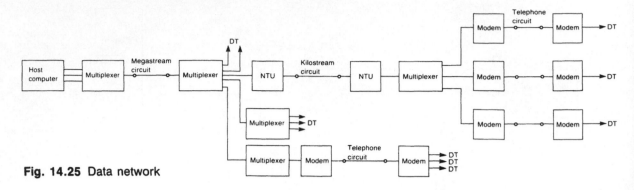

Fig. 14.25 Data network

Kilostream circuit nor a Megastream circuit is available at the location of some distant data terminals and then a network may use a combination of both analogue and digital circuits. Figure 14.25 shows a data network that includes both Megastream and Kilostream circuits as well as modems and analogue telephone circuits.

Exercises

14.1 Briefly explain the features of the BT services Kilostream, Megastream, Satstream and Switchstream.

14.2 Explain the method employed to transmit data over (*a*) a 64 kbits/s circuit and (*b*) a 2 Mbits/s circuit.

14.3 What is the difference between a concentrator and a multiplexer? Draw the block diagram of a data system in which two leased circuits and two links routed via the PSTN are connected to a multiplexer for transmission over a high-speed data link to another multiplexer which is connected to the front-end processor of a computer. Label each data link drawn with a typical bit rate.

14.4 List the factors which determine whether a leased point-to-point circuit or the PSTN should be used to connect a data terminal to a computer centre. Discuss how the decision will be affected by the use of (*a*) multiplexers, (*b*) concentrators.

14.5 Explain the need for point-to-point data links (*a*) between a computer and a data terminal and (*b*) between two computers.
    Show, with the aid of block diagrams, how two computers can be linked together when the distance between them is (*a*) very short, (*b*) a few kilometres, (*c*) 40 km.

14.6 Draw the block schematic diagram of a centralized computer network. Include examples of the use of both multiplexers and concentrators in your diagram. Why are cluster controllers often used?

14.7 Draw sketches to explain what is meant by the terms two-wire presented and four-wire presented when applied to presented data circuits. Give some reasons why a four-wire presented circuit is often used for data transmission.

14.8 Explain the need for point-to-point links in a data network. What are the advantages of leasing a private circuit? Give some possible ways in which a point-to-point link may be routed. Explain why modems are used on many long-distance data links.

14.9 Twelve 2400 bits/s data channels are simultaneously transmitted over a line using time-division multiplex. What is the bit rate on the line?

14.10 Six 2400 bits/s data links are connected to a concentrator which has three output links. What is the bit rate on each of the output links?

14.11 Discuss briefly the circumstances which may decide that a data link should be rented full-time and not dialled-up over the PSTN as, and when, required.

15 Integrated Services Digital Network

The public switched telephone network (PSTN) of the UK is now entirely digital for the transmission and switching of all trunk circuits. Transmission is over PCM systems operating at bit rates up to 2.4 Gbits/s and switching is carried out by digital exchanges using stored-programme control. This network is known as the *Integrated Digital Network* (IDN). In 1993 approximately 65% of the local telephone exchanges had been converted to digital operation but little progress had been made with the digitalization of the access network. When the access network has been converted to digital operation it will be known as the *Integrated Digital Access Network* (IDA). The combination of both the IDA and IDN will provide a multi-purpose digital telecommunication service that will be called the *Integrated Services Digital Network* (ISDN). ISDN is able to provide a user with an integrated set of voice and data services. The basic ISDN scheme is shown by Fig. 15.1.

BT offers two ISDN services, a basic rate service and a primary rate service. The basic rate service is intended for small business and home users and its service is known as either ISDN-2 or as 2B + D. 2B stands for two bearer circuits which are each allocated a separate number on the local telephone exchange, and D stands for a data circuit. The 2B + D service is shown in Fig. 15.1; it consists of two 64 kbits/s B channels and one 16 kbits/s D channel, a total of

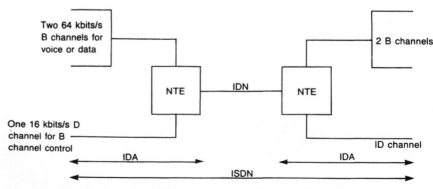

Fig. 15.1 Basic ISDN scheme (NTE = network terminating unit)

216

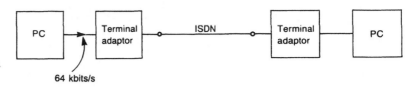

Fig. 15.2 Connecting two PCs via the ISDN

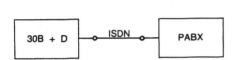

Fig. 15.3 30B + D circuit to a PABX

144 kbits/s. The 64 kbits/s channels may be used to carry either speech or data signals while the function of the D channel is to carry signalling information and to control the setting up and clearing down of calls set up on the two B channels. The D channel can also be used to transmit data. Figure 15.2 shows how a connection might be set up between two PCs over a link in the ISDN. Because ISDN lines are installed with two bearer circuits and a common control line their individual performance may be merged to give a single 128 kbits/s circuit. This merging process is known as *reverse multiplexing*.

The primary rate service is known as either ISDN-30 or as 30B + D; it provides the user with thirty 64 kbits/s speech or data channels plus a separate 16 kbits/s control channel. A primary rate circuit is intended for use by large businesses and to be connected directly to a digital PABX as shown by Fig. 15.3. The PABX is then used to switch lower bit rate circuits to their destinations. A primary rate ISDN service is also available from Mercury.

The use of the ISDN for the transmission of data has some advantages over the use of modems and the PSTN. The first advantage is that the bit rate made available is much higher at 64 kbits/s, and second is the much faster set-up time. On the other hand, most PCs are unable to transmit data at 64 kbits/s, since a fast UART like the 16550 is required.

Each ISDN customer will be offered a number of services including telephony, data, facsimile and video. The telephone service will provide more facilities than the existing service, such as (*a*) the diversion of incoming calls to another number if the called telephone is busy, (*b*) a busy telephone can be given an indication that another call is waiting, and (*c*) the number of each caller can be identified and displayed. Each customer will be provided with a switched data service that will not require the customer to have a modem. The applications of this service will include electronic mail, group IV facsimile, teletext, videotext and slow-scan television.

Appendix A
International Alphabet 5
or ASCII Code

The ASCII (American Standard Code for Information Interchange) codes for numbers, alphabet letters and other common symbols

Decimal numbers	ASCII code in binary	in hex	Alphabetical characters	ASCII code in binary	in hex
0	0110000	30	@	1000000	40
1	0110001	31	A (a)	1000001	41 (61)
2	0110010	32	B (b)	1000010	42 (62)
3	0110011	33	C (c)	1000011	43 (63)
4	0110100	34	D (d)	1000100	44 (64)
5	0110101	35	E (e)	1000101	45 (65)
6	0110110	36	F (f)	1000110	46 (66)
7	0110111	37	G (g)	1000111	47 (67)
8	0111000	38	H (h)	1001000	48 (68)
9	0111001	39	I (i)	1001001	49 (69)
			J (j)	1001010	4A (6A)
Other			K (k)	1001011	4B (6B)
symbols			L (l)	1001100	4C (6C)
:	0111010	3A	M (m)	1001101	4D (6D)
;	0111011	3B	N (n)	1001110	4E (6E)
<	0111100	3C	O (o)	1001111	4F (6F)
=	0111101	3D	P (p)	1010000	50 (70)
>	0111110	3E	Q (q)	1010001	51 (71)
?	0111111	3F	R (r)	1010010	52 (72)
Space	0100000	20	S (s)	1010011	53 (73)
!	0100001	21	T (t)	1010100	54 (74)
"	0100010	22	U (u)	1010101	55 (75)
#	0100011	23	V (v)	1010110	56 (76)
$	0100100	24	W (w)	1010111	57 (77)
%	0100101	25	X (x)	1011000	58 (78)
&	0100110	26	Y (y)	1011001	59 (79)
'	0100111	27	Z (z)	1011010	5A (7A)
(	0101000	28			
)	0101001	29	[	1011011	5B
*	0101010	2A	\	1011100	5C
+	0101011	2B	]	1011101	5D
,	0101100	2C	↑	1011110	5E
-	0101101	2D	←	1011111	5F
.	0101110	2E	a	01100001	61
/	0101111	2F	b	01100010	62

Decimal numbers	ASCII code in binary	in hex	Alphabetical characters	ASCII code in binary	in hex
			c	01100011	63
			d	01100100	64
			e	01100101	65
			f	01100110	66
			g	01100111	67
			h	01101000	68
			i	01101001	69
			j	01101010	6A
			k	01101011	6B
			l	01101100	6C
			m	01101101	6D
			n	01101110	6E
			o	01101111	6F
			p	01110000	70
			q	01110001	71
			r	01110010	72
			s	01110011	73
			t	01110100	74
			u	01110101	75
			v	01110110	76
			w	01110111	77
			x	01111000	78
			y	01111001	79
			z	01111010	7A

Answers To Numerical Exercises

Chapter 1 **1.6** (a) Bit rate = $1/(104.17 \times 10^{-6})$ = 9600 bits/s
(b) Baud speed = 9600
1.7 (a) Bit duration = 1.667/4 = 0.416 75 ms
Baud speed $1/(0.416\,75 \times 10^{-3})$ = 2400
(b) Baud speed = 2400
1.8 Bit duration = 1/2400 = 416.67 μs
f = 1200 Hz
1.11 (a) Baud speed bit rate = 14 400
(b) Bit duration = 1/14 400 = 69.4 μs

Chapter 2 **2.1** $v = f\lambda = 50 \times 10^3 \times 5200 = 2.6 \times 10^8$ m/s
2.2 $\lambda = (3 \times 10^8)/(300 \times 10^6) = 1$ m, $\lambda/4$ = 0.25 m
2.3 $\lambda = (3 \times 10^8)/60 \times 10^3) = 5000$ m
2.4 Maximum frequency change = $10 \times 100 = 1000$ Hz
2.5 Tolerance = 450/90 = ±5 ppm
2.6 $\lambda_1 = (3 \times 10^8)/(1 \times 10^6) = 300$ m
$\lambda_2 = (3 \times 10^8)/(1.015 \times 10^6) = 295.57$ m
Separation = $300 - 295.57 = 4.43$ m

Chapter 3 **3.6** (a) One carrier cycle in 0.5/5 = 0.1 ms
$f_c = 1/(0.1 \times 10^{-3}) = 10$ kHz
(b) One envelope cycle in 0.5 ms
$f_m = 1/(0.5 \times 10^{-3}) = 2$ kHz
(c) V_{min} = 25 V, V_{max} = 100 V
$m = (100 - 25)/(100 + 25) = 0.6$
3.7 f_c = 72 kHz, $f_l = 72 - 2 = 70$ kHz, $f_u = 70 + 2 = 72$ kHz
3.8 $P_t = 10 = P_c(1 + 0.6^2/2)$
P_c = 8.475 kW
$P_{sf} = 10 - 8.475 = 1.525$ kW
3.9 f_c = 500 kHz
LSB = 500 kHz $-$ (250 to 3000 Hz) = 497 to 499.75 kHz
USB = 500 kHz $+$ (250 to 3000 Hz) = 500.25 to 503 kHz
Bandwidth = 6 kHz
3.11 f_c = 50 kHz, f_l = 47.5 kHz, f_u = 52.5 kHz, Bandwidth = 5 kHz
m = 6/12 = 0.5
3.12 (a) LSB = 480 to 490 kHz (b) USB = 510 to 520 kHz
Bandwidth = $520 - 480 = 40$ kHz

3.13 $V_{max} = 28 \times 1.5 = 42\,V$
$V_{min} = 28 \times 0.5 = 14\,V$

3.14 $V_c + V_m = 12$, $V_c - V_m = 3$
Subtracting, $2V_m = 9$ and $V_m = 4.5\,V$
Adding, $2V_c = 15$ and $V_c = 7.5\,V$
$V_{USF} = 4.5/2 = 2.25\,V$

3.15 $V_1 = 6 = V_m/2 \Rightarrow V_m = 12\,V$
$m = 12/24 = 0.5$

3.16 $P_{sf} = 100 \times 0.4^2/2 = 8\,W$
$P_{lsf} = 4\,W$

3.17 $P_t = 1000(1 + 0.64/2) = 1320\,W$

3.18 $m = 0.3/3 = 0.1 = 10\%$

3.20 $12 = 9(1 + m)$, $m = 12/9 - 1 = 0.33$
$V_s = (0.33 \times 9)/2 = 1.5\,V$

3.21 $P_c = (9/\sqrt{2})^2/100 = 405\,mW$
$P_t = 405(1 + 0.33^2/2) = 427\,mW$

3.22 $LSB_A = 88 - (0\ to\ 4) = 84\ to\ 88\,kHz$
$LSB_B = 120 - (84\ to\ 88) = 32\ to\ 36\,kHz$

3.23 (a) $P_t = 2000 + 100 + 100 = 2200\,W$
(b) $2200 = 2000(1 + m^2/2)$, $m = 0.447$

3.24 $V_{max} = 0.01 + 0.001 = 0.011\,V = 11\,mV$

3.26 (a) $V_m = 0.25 \times 50 = 12.5\,V$
(b) $V_{sf} = 12.5/2 = 6.25\,V$
(c) $V_{LSF} = V_{USF} = 6.25\,V$

3.27 (a) Frequency swing $= 2 \times 10 = 20\,kHz$
(b) $10\,kHz$

3.28 $B = 2(6 + 3) = 18\,kHz$
Number of channels $= 500/22 = 22$

3.29 $B = 2(7 + 4) = 22\,kHz$

3.30 (a) $m_f = 70/15 = 4.67$
(b) Frequency swing $= 2 \times 70 = 140\,kHz$

3.31 (a) $B = 150 = 2(f_d + 20)$, $f_d = 55\,kHz$
$D = 55/20 = 2.75$
(b) Frequency swing $= 2 \times 55 = 110\,kHz$

3.32 (a) $D = 2 = 6/f_{m(max)}$, $f_{m(max)} = 3\,kHz$
(b) $B = 2(6 + 3) = 18\,kHz$

3.33 $100\,W$

3.34 $kf_d = 20 \times 0.5 = 10\,kHz$

3.37 $6 = f_d/10$, $f_d = 6 \times 10 = 60\,kHz$

Chapter 4 **4.2** (a) $f_{max} = 600/2 = 300\,Hz$
(b) (i) minimum bandwidth $= 600\,Hz$
(ii) for third harmonic $= 1800\,Hz$

Chapter 6 **6.1** Loss $= 10\sqrt{4} = 20\,dB$
6.2 Loss at $4\,MHz = 4\sqrt{(4/2)} = 5.657\,dB$
Loss of $2.5\,km = 2.5 \times 5.657 = 14.14\,dB$
6.3 Input current $= 50/1200 = 41.67\,mA$
Loss $= 1 \times 20 = 20\,dB$, $20 = 20\log_{10}(41.67/I_r)$
$10^1 = 10 = 41.67/I_r$, and $I_r = 4.167\,mA$

6.4 $I_{in} = 10/1200 = 8.33 \, \text{mA}$, $P_1 = 5 \times 10^{-3} = I_1^2 \times 600$
$I_1 = \sqrt{[(5 \times 10^{-3})/600]} = 2.89 \, \text{mA}$
Loss $= 20 \, \log_{10}(8.33/2.89) = 9.2 \, \text{dB}$
Length of line $= 9.2/0.8 = 11.5 \, \text{km}$

6.6 $Z_{in} = Z_0 = 75 \, \Omega$

6.7 $V_g = \delta\omega/\delta\beta = [2\pi(15 - 5) \times 10^3]/(0.419 - 0.175)$
$= 257.5 \, \text{km/s}$

6.8 Loss $= 20 \, \log_{10}(12/4) = 9.54 \, \text{dB}$
$= 9.54/4 = 2.39 \, \text{dB/km}$

6.9 Loss $= 1 \times 6 = 6 \, \text{dB at 1 kHz}$
$= 1.8 \times 6 = 10.8 \, \text{dB at 3 kHz}$
6 dB = voltage ratio of 2.
Hence, V_{out} at 1 kHz $= 10/2 = 5 \, \text{V}$
10.8 dB = voltage ratio of 3.47.
Hence, V_{out} at 3 kHz $= 2.5/3.47 = 0.72 \, \text{V}$

6.10 $S = 20/4 = 5$

Chapter 7

7.1 $\varphi_{r(min)} = \sin^{-1}(1/1.68) = 36.53°$

7.6 $\varphi_c = \sin^{-1}(0.99\eta/\eta) = 81.9°$

7.7 $1/(1 \, \mu s) = 1 \, \text{MHz}$
Maximum bit rate $= 1/(2 \, \mu s) = 500 \, \text{kbits/s}$

7.10 (a) $2r_1 = 50 \, \mu m$
$= 50/0.85 = 58.8\lambda$
(b) $2r_1 = 5 \, \mu m = 5/0.85 = 5.88$

7.11 (a) $(2 \times 10^9)/(20 \times 10^3) = 100 \, \text{kHz}$
(b) $(2 \times 10^9)/(500 \times 10^6) = 4 \, \text{km}$

Chapter 8

8.1 $V_n = \sqrt{(4 \times 1.38 \times 10^{-23} \times 293 \times 10^5 \times 78 \times 10^3)}$
At the output $V_n = 11.23 \, \mu V \times 150 = 1.69 \, \text{mV}$

8.3 $V_n = \sqrt{(4 \times 1.38 \times 10^{-23} \times 290 \times 2 \times 10^6 \times 500 \times 10^3)}$
$= 126.5 \, \mu V$

8.4 $kTB = 1.38 \times 10^{-23} \times 291 = 4 \times 10^{-21} \, \text{W/MHz}$

8.6 Signal-to-noise ratio $= 20 \, \log_{10}(10/0.1)$
$= 40 \, \text{dB}$

8.7 $37 = 10 \, \log_{10}[S/(100 \times 10^{-6})]$
$10^{3.7} = 5012 = S/(100 \times 10^{-6})$, and $S = 0.5 \, \text{W}$

8.8 $kTB = 1.38 \times 10^{-23} \times 290 \times 100 \times 10^3 = 4 \times 10^{-16} \, \text{W}$

8.9 $S_{in} = -10 \, \text{dBm}$; Gain $= +17 -(-10) = 27 \, \text{dB}$

8.10 45 dB $= 31\,623$ power ratio
$S_{in} = 31\,623 \times 1.38 \times 10^{-23} \times 290 \times 2 \times 10^6 = 253 \, \mu W$

Chapter 9

9.7 4 bits occupy 0.833 ms
$T = 2 \times 0.833 = 1.666 \, \text{ms}$
$f = 1/(1.666 \times 10^{-3}) = 600 \, \text{Hz}$

9.8 (a) $f_{min} = 0 \, \text{Hz}$
(b) $f_{max} = 4800/2 = 2400 \, \text{Hz}$

9.10 Bit rate $= 1/(1.042 \times 10^{-4}) = 9600 \, \text{bits/s}$

Chapter 10

10.2 Zero

10.3 20 to 25 kHz

10.8 12–15 kHz

10.11 9 dB: $N = 10^{0.45} = 2.82$

$R_1 = R_3 = 140(2.82 - 1)/3.82 = 66.7 \simeq 68\,\Omega$

$R_2 = (2 \times 140 \times 2.82)/(2.82^2 - 1) = 113.6 \simeq 110\,\Omega$

10.12 $N = 2.82$

$R_1 = R_3 = 140(2.82^2 - 1)/(2 \times 2.82) = 172.6 \simeq 180\,\Omega$

$R_2 = 140(2.82 + 1)/(2.82 - 1) = 293.8 \simeq 300\,\Omega$

10.13 Attenuator loss $= (-7 + 27) - 10 = 10\,dB$

10.14 $N = \log_{10}^{-1}(15/20) = 5.62$

$R_1 = [75(5.62 - 1)]/(5.62 + 1) = 52\,\Omega$

$R_2 = (2 \times 75 \times 5.62)/(5.62^2 - 1) = 28\,\Omega$

10.15 $N = \log_{10}^{-1}(20/20) = 10$

$R_1 = 600(10^2 - 1)/(2 \times 10) = 2970\,\Omega$

$R_2 = 600(10 + 1)/(10 - 1) = 733\,\Omega$

Chapter 12

12.8 (a) $2^n = 256$, $n = 8$

(b) $2^n = 8192$, $n = 13$

12.10 $1024 = 2^{10}$

Sampling period $= 1/(15 \times 10^3) = 66.67\,\mu s$

Each sample occupies $666.7\,\mu s$

Number of pulses $= 1/(666.7 \times 10^{-6}) = 1500$

12.11 Each sample occupies $8 \times 0.488 = 3.9\,\mu s$

Intervals between samples $= 1/8000 = 125\,\mu s$

Number of channels $= 125/3.9 = 32$

12.12 Line bit rate $= 8 \times 8000 \times 32 = 2.048\,Mbits/s$

For 8192 levels, bit rate $= 13 \times 8000 \times 32 = 3.328\,Mbits/s$

12.15 0 V, 8.66 V, 8.66 V

0 V, -8.66 V, -8.66 V

12.16 $0.54/0.01 = 54$

12.17 Level size $= V_{max}/128$ or $V_{max}/512$

% reduction $= [V(1/128 - 1/512)/(V/128)] \times 100 = 75\%$

Chapter 14

14.9 Line bit rate $= 12 \times 2400 = 28.8\,kbits/s$

14.10 2400 bits/s

Index